T0361848

"3D printing has become emblematic for deep-seated, ambivalent, and strange changes in our societies. Increasingly accessible, and increasingly powerful, it is unclear whose digital fabrications will be served by this potentially ubiquitous technology. Will it be manufacturers, hackers, makers, the military, peer-producers, or entirely new social figures? And what about relations between capital, labour, consumption and environment? Birtchnell and Urry provide a clear-sighted and measured analysis into these issues. Drawing upon the historical, geographic and social relations shaping the development of this technology, their book navigates some of the futures open to us, and makes clear the social choices involved right now."

**Adrian Smith**, *Professor of Technology and Society,*
*University of Sussex, UK*

# A NEW INDUSTRIAL FUTURE?

*A New Industrial Future?* examines whether a further industrial revolution is taking place around the world. In this compelling book Birtchnell and Urry examine such a new possible future involving the mass adoption of 3D printing. The locating of 3D printers in homes, offices, stores and workshops would disrupt existing systems and pose novel challenges for incumbents. The book – drawing upon expert interviews, scenario workshops and various case studies – assesses the potential future of global manufacturing, freight transport, world trade and land use. It offers the first book-length social scientific analysis of the character and impacts of a new system of manufacturing that is in formation. The book will be of interest to urban planners, policy makers, social scientists, futurologists, economists, as well as general readers by offering inquiry on this future upheaval in the means of production.

**Thomas Birtchnell** is a Senior Lecturer in Geography and Sustainable Communities at the University of Wollongong, Australia.

**John Urry** was Distinguished Professor of Sociology and Co-Director of the Institute for Social Futures, Lancaster University, UK.

ANTINOMIES
# Innovations in the Humanities, Social Sciences and Creative Arts
*Series Editors: Anthony Elliott and Jennifer Rutherford*
*Hawke Research Institute, University of South Australia*

This series addresses the importance of innovative contemporary, comparative and conceptual research on the cultural and institutional contradictions of our times and our lives in these times. *Antinomies* publishes theoretically innovative work that critically examines the ways in which social, cultural, political and aesthetic change is rendered visible in the global age, and that is attentive to novel contradictions arising from global transformations. Books in the series are from authors both well-established and early careers researchers. Authors will be recruited from many, diverse countries – but a particular feature of the series will be its strong focus on research from Asia and Australasia. The series addresses the diverse signatures of contemporary global contradictions, and as such seeks to promote novel transdisciplinary understandings in the humanities, social sciences and creative arts.

The Series Editors are especially interested in publishing books in the following areas that fit with the broad remit of the series:

- New architectures of subjectivity
- Cultural sociology
- Reinvention of cities and urban transformations
- Digital life and the post-human
- Emerging forms of global creative practice
- Culture and the aesthetic

Series titles include:

1. **The Consequences of Global Disasters**
   *Edited by Anthony Elliott and Eric L. Hsu*

2. **A New Industrial Future? 3D printing and the reconfiguring of production, distribution, and consumption**
   *Thomas Birtchnell and John Urry*

# A New Industrial Future?

3D printing and the reconfiguring of production, distribution, and consumption

*Thomas Birtchnell and John Urry*

Routledge
Taylor & Francis Group

LONDON AND NEW YORK

First published 2016
by Routledge
2 Park Square, Milton Park, Abingdon, Oxon OX14 4RN

and by Routledge
711 Third Avenue, New York, NY 10017

*Routledge is an imprint of the Taylor & Francis Group, an informa business*

© 2016 Thomas Birtchnell and John Urry

*British Library Cataloguing-in-Publication Data*
A catalogue record for this book is available from the British Library

*Library of Congress Cataloging in Publication Data*
Names: Birtchnell, Thomas, author. | Urry, John, author.
Title: A new industrial future? : 3D printing and the reconfiguring
of production, distribution, and consumption / by Thomas Birtchnell
and John Urry.
Description: 1 Edition. | New York : Routledge, 2016.
Identifiers: LCCN 2016002023| ISBN 9781138022911 (hardback) |
ISBN 9781138022928 (pbk.) | ISBN 9781315776798 (ebook)
Subjects: LCSH: Manufacturing industries–Technological innovations. |
Printing industry–Technological innovations. | Information technology–
Economic aspects.
Classification: LCC HD9720.5 .B547 2016 | DDC 338/.064–dc23
LC record available at https://lccn.loc.gov/2016002023

ISBN: 978-1-138-02291-1 (hbk)
ISBN: 978-1-138-02292-8 (pbk)
ISBN: 978-1-315-77679-8 (ebk)

Typeset in Bembo
by Cenveo Publisher Services

# CONTENTS

# LIST OF FIGURES

# FOREWORD

When I commissioned this book at Routledge for the *Antinomies* series, I had little idea that *A New Industrial Future?* would be the last collaborative book that John Urry would write – along with his protégé and coauthor Thomas Birtchnell. It is interesting that *A New Industrial Future?*, alongside John's recent *What is the Future?*, appear as titles with question marks. This, I would suggest, reflects John's uncanny ability to raise thorny questions at just the right political time – something he did most startlingly with the publication of *Offshoring* in 2014, anticipating the global significance of financial outsourcing and tax evasion as captured most recently by the Panama Papers leak of 2016. John was a sociological thinker to his core, and it was social critique (of the kind practiced from the Frankfurt School of the 1920s to Zygmunt Bauman in our own time) that he did in spades and to brilliant effect.

John Urry's own brand of social theory took off in the 1980s and 1990s, initially with his collaborations with Scott Lash on labour and capitalism, and subsequently bloomed with full force in the early 2000s with his research on mobilities and globalization. John captured the complex, contradictory ways in which the advent of the global electronic economy is intricately intertwined with movement, fluidity, and the wholesale transformation of transport, travel and tourism. Urry's 'mobilities paradigm' had enormous impact in the UK especially, and in various parts of Europe and Australasia too. Understanding the breadth and depth of "life on the move" for institutions, organizations, regions, nation-states and indeed global governance was at the core of John's intellectual project. But so too was the impact of mobilities on professional and personal life, and it was my good fortune to have worked with John on our book *Mobile Lives* (Routledge) at the turn of the 2010s.

In *A New Industrial Future?*, Urry and Birtchnell assess the systemic implications of alternative ways of organizing, disorganizing and reorganizing world systems of trade and ground their ideas in recent innovations and developments in the area of 3D printers. Just as Urry had critically interrogated the motorcar in his edited collection *Automobilities*, co-edited with Nigel Thrift and Mike Featherstone, and mobile phones in *Mobile Technologies of the City*, co-authored with Mimi Sheller, the 3D printer was something that provoked his critical attention.

Urry was a rare bird in sociology, not least of which because – unlike many social scientists - his ideas were of direct relevance to non-academics, ranging from environmentalists to policymakers. He contributed to the Intergovernmental Panel on Climate Change and their work on transportation. He was also appointed to the UK Government's Foresight Program on transport and policy futures, which in turn led to his vital research on social futures as well as his setting up an Institute for Social Futures during what would be the final years of his life at Lancaster University.

More than all of this, however, is the sheer novelty of John's take on the world, the angles with which he could make the ordinary appear extraordinary, with all sorts of social, cultural, political and environmental consequences of which had rarely been previously addressed. As I have said, he had the rare knack of being able to address both scholarly and lay audiences alike. Above all, he was a gentle and generous man. It is with great sadness for my friend and colleague that I write this Foreword, but heartening too that his final posthumous book now sees the light of day, with a work wholly testament to a mind untrammelled by the conventions of academic fashion.

*Anthony Elliott*
*Adelaide, 2016*

# PREFACE

Sadly, John Urry, my co-author, mentor and friend, passed away on the 18th March this year before seeing this book come to print. He had just finished a prestigious tour of China and was preparing for a talk in London at University College London (UCL). On the 18th we shared correspondence where he voiced his satisfaction at receiving the proofs for this book at the same time as his forthcoming book *What is the Future?* (Polity). He also urged me to begin this Preface. Characteristically of John, the acknowledgement of his rich professional and social networks was at the forefront of his mind. He was always at pains to give thanks where they were due.

My path crossed John's at the end of his career, but it did not feel like he was slowing down. On the contrary, in my three years working with John at Lancaster University he published three field-defining books amidst much other activity. The launch of his new centre, the Institute for Social Futures, seemed to suggest an even more busier and fruitful future for him to come.

The above achievements for John appeared alongside our day-to-day project work for the Economic and Social Research Council (ESRC) with our colleagues at the excellent Centre for Transport and Society, University of the West of England (UWE), and generous supporters in the UK Transport Research Centre. It was in this project that we conceived of the ideas presented in this book, mostly on train stations and in conferences or chatting during cafe, food cart, lunch hall and restaurant meetings. John was always on the move.

I recall sitting outside for our weekly lunch in Lancaster's cold sunlight in the food-court at the university enjoying a spicy curry. I had requested a time-consuming mango lassi and John had his usual filter coffee. We would often reflect, on these occasions, on our mutual interest in technology adoption and we discovered that day that we had both come to rest upon the topic of 3D printing in our separate readings in the press.

Finding this common ground we mused on integrating 3D printing into our project as a non-transport technology which could impact significantly upon transport in the future. Neither of us really cared particularly about the technology itself (we were not rushing out to buy one ourselves), but rather the kinds of future worlds and societies it made it possible to imagine.

The spectre of technological determinism loomed large over our discussions on 3D printing as did John's close colleagues working in this area. A key feature of the ideas in this book is both the early and mass adoption of technology and the hype that accompanies innovation. John and I shared a background of reading in both actor network theory and complex systems theory. We often mused on technological agency. John and I pondered in our meetings about the nature of choice in how technologies come to be adopted en masse as 'socio-material systems'; however, we also discussed the way in which technologies could be said to adopt people too in terms of entanglement in their lives. For instance, Latour's example of an automatic door-closer regulating *which* humans can enter a room and which cannot (e.g., children or the wheelchair bound).

Unlike me, John was a reluctant adopter of technologies; however, he also understood the necessity of using them for work and pleasure and indeed made much use of texting and email on his smart phone. I recall him enthusing about his hybrid electric car: its fuel efficiency and lack of noise.

To my mind, we were both keenly interested in how, in many cases, systems are designed so that people have very little choice about whether they adopt technologies or not. Instead of being 'early adopters' of technology (as many marketeers incite people to be) most are anxious about technology adoption. It could be said that systems, and the technologies which are a part of them, adopt people. For instance, in an automobile-dominated world people without cars are unable to access social networks or basic services or to go shopping. Similarly, without a mobile phone people run the risk of being imperilled in emergencies, or isolated from their family, or stranded.

So we both liked to think (perhaps cryptically) that users are adopted rather than adopters: there are 'automobile adoptees', 'smart phone adoptees', 'automatic teller machine adoptees' and so on. Adoption is part and parcel of being participants in systems and similar to being adopted into a family, perhaps. There is agency and indeed power at work between humans and non-humans. In this way path dependencies and 'lock-ins' (as John's peer Brian Arthur put it) matter in the future adoption of people by the 3D printing system. I am not sure if John would agree with this idea or not on paper, but it certainly has underpinned my own thinking in this book.

John was a tireless networker and I found his true métier was in convincing people in government and business that social science was in fact serious business. In our many meetings, interviews and presentations with people in policy and industry on this topic, I was struck by the way John could impress upon professionals and policymakers the gravity of his research and ideas. He was adept at bringing together and making feel comfortable 'need-to-meet' people from disparate places.

There is no question that John was a maverick himself (I remember my delight when I discovered he once began an article in the highly ranked journal *Global Environmental Change* with a quote from a Joni Mitchell song). John could afford to be a maverick because of his cognizance of the systems of academia and government and his intuition of how to influence them. He always had time for free-thinking intellectuals who asked uncomfortable questions which placed them outside of the hallowed halls. Indeed he thrived off of them and brought them into the fold. His networks in the academy were immense and he made equal time for both the eminent and the early in career.

No other social scientist has had the bravery to confront as many big issues as John. I am sure in decades and even centuries to come his legacy will be his own program of research alongside the knowledge products of his energetic collaborations. I will remember John for his humility, sense of humour and ease in company. I was always struck by his pride in Lancaster University and its Sociology department.

In times to come I hope John's efforts to draw attention to the plight of the environment, the hollowness of high-carbon living, the calamity of climate emissions, and the multiplicity of social worlds we might live in are a beacon for real change. One recent scholar styled John as the 'harbinger of the death of distance'; however, to my mind he is better styled as a companion and guide for all of us to a more equal, cleaner, communal and considerate future.

Many people assisted with the formation of this book, unfortunately too many for me to name individually without John's aid. From the onset of this book project we intended to thank colleagues at Lancaster University and the University of Wollongong (UOW) for their collegial support and input. We also acknowledge the financial support of the ESRC (ES/J007455/1) and Australian Research Council (ARC) Discovery Project (DP160100979) and our partners in these grants: Glenn Lyons (UWE), Christa Hubers (Tilburg), Peter Jones (UCL), Anthony Elliott, (University of South Australia) and David Bissell, (Australian National University). Our colleagues from the publisher, Gerhard Boomgaarden and Alyson Claffey, deserve much credit for their commitment to John's work throughout Routledge's history with him and their patience with this particular project. So too do the many people who gave John and me their precious time for our workshops and interviews. We hope we do credit to their work. We especially thank Robert Gorkin from UOW in particular for his assistance as our consultant on the technical aspects of 3D printing throughout the book and for his leading the production of Figure 6.2. Thanks also to Joelle Dietrick for allowing us to use her amazing art for the cover and featured artwork (Figure 6.1) and Gil Viry for his exceptional social network analysis visualizations and collaboration with us (Figure 2.1).

John once shared with me that his love of books came from his time spent in the now famously eccentric Foyles Bookstore on Charing Cross Road in London. I can only speculate that it would have given him much satisfaction to see his own name appearing yet again on those particular shelves as a popular academic author.

# 1

# THINKING ADDITIVELY

## Introduction

Industry consultants Wohlers Associates foresee the worldwide market for 3D printing products and services growing to US$10.8 billion by 2021 and driving economic growth.[1] Industry consultants Gartner are more cautious in their foresight, noting a wide degree of hype with many hurdles to mass adoption by consumers and producers.[2] Already estimates are being made that the global 3D printing market will reach approximately US$3 billion by 2018 according to the executive summary of the report '3D Printing – A Global Strategic Business Report' by *Global Industry Analysts*.[3] The decentralization and distribution of the means of production through digital fabrication could be epochal since it involves a reconfiguration of the current systems transforming the very notion of manufacturing. Anticipated changes are a closer proximity of production to design as consumers access online repositories of files to print out themselves. A forerunner is the innovation of the Internet whereby music or films or books became downloadable at a price and in some cases for no cost at all.[4]

Three decades ago 3D printers became available in the pre-production, or prototyping, stages of manufacturing objects. Architects, industrial engineers, designers and other users requiring models for testing and bringing concept-to-prototype now use 3D printers every day. When early forms of 3D printing, known as 'stereolithography apparatus' (SLA), became available in the mid-1980s, there was already speculation that it would be a game-changer for production systems. The inventor of this early process and founder of major 3D printing company 3D Systems, Charles 'Chuck' Hull, cited a report by Californian research firm DATAquest describing how the innovation works. He noted: 'Stereolithography has the potential to change the (manufacturing) industry, as we know it today. Never before has a manufacturing process achieved such dramatic time and expense reduction in prototypic manufacturing'.[5] SLA was integrated into pre-production processes for the purpose of what became known as 'rapid prototyping', whereby a user could produce an experimental or unfinished object for testing without having to resort to costly factory settings, which are only economically feasible for very many finished products. Alongside polymers other heat sensitive materials became popular as rapid prototyping went mainstream.

A visitor to the London 3D Print Show will realize that the major interests in 3D printing involve rapid prototyping either through bureaus or desktop 3D printers, yet this is rapidly changing. Many experts are examining both the wider implications of 3D printing for economies and societies and how the social world itself will structure the consequences of 3D printing.[6] It should be noted that there are many different processes here: heated extrusion, laser sintering, electron beam melting, chemical binding and others still in testing. Consumers can purchase more affordable versions off the shelf in major high street suppliers; or make use of services with industrial machines and professionally trained staff.

Various innovations in scanner, sensor, laser, electron beam and chemical technologies have been pivotal in bringing these processes into product development and allowing certain 3D printers to reach the market. As the technologies grouped under the umbrella term '3D printing' have matured into consumer and industrial products they look more like 2D printers, whether this is the latest desktop unit or the larger model of 'office' type printer. However, there will not be a like-for-like emergence. Indeed 3D printers could combine with other technologies into a 'digital fabricator', such as suggested by the Kickstarter project 'Maker-Arm': a complete digital fabrication system combining a wirelessly controlled robotic arm with a 3D printer, laser-cutter, carver, printed circuit board (PCB) mill, drill, solder-paste dispenser, pick and place assembly, and automated solderer.[7] 3D printing is just one of many automation machines.

It should also be noted that 3D 'prints' do not appear out of thin air. Each object requires feedstock as well as some additional material for scaffolding, and in some models of printer enough excess material to fill the build tray to capacity. Feedstock refers to stock material that is fed into a printer, whether in the form of powder, liquid, gel, filament, gas or some other standardized raw material.[8] Initial innovations in the printing of objects occurred in experiments with lasers, chemicals and binding agents and were a progression from the moulding of materials to change their state through a phase shift. Rather than pouring or injecting a material into a mould (as in the modern process of 'injection moulding' and in the practice of metal casting where metal is melted and poured into a mould) a computer solidifies material feedstock layer by layer with micro-millimetre detail. Because of the visibility of the layering in many cases of 3D printing, some form of minor 'finishing' is often required for objects to be visually satisfactory.

In low-end 3D printing an object is produced using affordable and widespread composite plastic filament (known as ABS or acrylonitrile-butadiene-styrene) through a process of melting, extrusion and solidification. This innovation made 3D printing affordable and also freed the process from control by material patents and expensive printer technologies. And more importantly, with the use of an end product material such as ABS the idea of 'rapid manufacturing' became a possibility. With open source and commercial desktop laser sintering this likelihood increases markedly.[9]

Projected market barriers for 3D printing are the total cost (including materials, software, designs, power), the usability of software (interface and design), production times and product quality.[10] Objects now 3D 'printed' include many plastic consumer accessories and novelties; metal car, aeroplane and motorbike parts; textiles, ornaments and clothing; and even cardiovascular tissue.[11] In some cases, nature inspires the designs now made possible through this technique, for instance honeycomb structures in bioengineering.[12] Alongside the diversification of objects being printed there are also a growing number of websites offering repositories of designs and online services in a range of possible materials. These online services range from open source and peer-to-peer aggregators of user-submitted designs to multinational businesses with sophisticated supply chains and products ranging from household objects and novelties to luxury items.

In this book we are suggesting that in the future there will be far more flexibility, choice and variety in the global system people currently rely on for the objects they use in everyday life. As a consequence social, geopolitical and economic upheaval, disruption and transformation are also on the horizon. The ubiquity of the automobile, personal computer or smart phone certainly suggests there could be social ramifications for 3D printing. However, a closer and critical examination of these technologies and others suggests that combinations of different innovations will result in a reconfiguration of the incumbent systemic triad of production, distribution and consumption.[13] At the moment predictions of the social impacts of 3D printing take into account the range of printers and the power and significance of existing interests, including patterns of low cost manufacturing, containerization and supply chains, and brand-centric capitalism. We will see how 3D printing might be transformative of very many freight miles and this requires examination of contrasting scenarios for 2050 and the forms in which novel socio-technical systems might emerge and generate different forms of personal and object transportation in coming decades. We consider 3D printing through examining its potential to reconfigure the triad.

3D printing is not completely unforeseen or left field. There are already additive manufacturing systems in place in today's society. The conventional retail high street includes both suppliers of manufactured objects, those who craft or repair an already manufactured item, and ones who combine various prefabricated ingredients into a product on site. A shopper might choose to have their shoes re-heeled by a high street cobbler (an additive process of repair, as replacement soles are layered on top of the existing one), and while they wait will buy something to eat from a donut stall where an operator uses an extrusion machine to combine a mixture of dough, sugar and other ingredients into a product deep-fried in a second machine where it is 'finished': glazed with sugar or icing.

A common sight in the world's cities, the donut seller, uses a 3D printer-like extrusion technology. Certainly, the ingredients for the donut-maker's wares are likely to have been transported some distance from farming and packaging facilities. The economic and logistical benefits of extruding donuts on site stem from profits made in standardizing and optimizing the weight, density and packaging of the ingredients, which keeps overhead costs down. And the donut seller can also advertise the donuts as 'freshly made' regardless of the distance and time between origin and sale.

Similarly, coffee shops are places for meeting people, spending time working, as well as purchasing hand-made coffee. A popular iPhone app called 'London Coffee' illustrates the growing success of boutique cafés in central London with an impressive listing of small businesses. Such cafés will have an espresso machine – that is, another heat extruder – alongside other types of coffee equipment. Feedstock materials for the espresso machine include a range of coffee beans for processing in a blender, which converts them into a powder. A successful café involves not only stocking the best coffee but also training the most proficient 'baristas'.

Baristas pay attention to the cleanliness of their equipment, the length of time they heat milk, the angle of the frothing wand, the consistency of the crema and the movement of the jug as they pour milk to make 'marbling' effects on the surface of the coffee: their craft. There is an acknowledgement of artistry in coffee preparation and baristas receive accredited training, with some going on to compete in regional, national and international competitions, where they receive additional credit for their skills. Being a barista can be a lifetime career combining technical skill, presentation and a degree of 'flair'. The growth of boutique cafés alongside a wide range of affordable home espresso machines is a useful analogy for 3D printing.

Similar to coffee making, 3D printing necessitates a material feedstock; loose powders or cartridges; a technical process involving blending, heating and extrusion; and some technical skill in making all this work together. And like 3D printing there are high-end coffee machines for industry purposes and low-end home units too. There is a wide range of product quality types and consumer expectations.

So what do these food and beverage industries suggest for the future of 3D printing? First, the widespread proliferation of home or desktop 3D printers would not necessarily spell disaster for industry, craft or bureau 3D printing. While it is true that innovations in digital printing saw the demise of the ubiquitous camera film print shop, there remains niche demand for digital print shops where consumers can print out copies of their digital photos. Unlike the camera film industry, the café industry has grown alongside the development of a huge market in highly technical and affordable home espresso machines. Furthermore, the home printing of exotic materials such as steel, titanium, aluminium and ceramics will require innovations in printing technology currently inconceivable by today's standards. These materials require expensive high-end machines with technically trained users. Such rumination is relevant to the future of food products specifically, with one recent study on the chocolate industry highlighting that the adoption of 3D printing technology and the mass customization it affords is 'a matter of survival' for manufacturers.[14] This chapter thus examines the pending system innovation arising from the digitization of data and the decentralization of 3D printers engendering a new system that will be unfamiliar to people today.

## From 2D to 3D

The subject of this book might be a singularly known technology – that is, 3D printing – however, we wish to distance ourselves from deterministic accounts of social change arising from this innovation. There is a way to imagine 3D printing for everyday, household objects becoming ubiquitous without sinking into deterministic thinking, namely that it would arise from a transformation in an epochal sense of what we understand as 'capitalist' societies.

Elaboration on a new system driving the ubiquity of 3D printing is far from whimsical. With the imminent threats to society of climate change governments around the world are now assessing the future of high-carbon living and the greenhouse gas emissions (GHGs) occurring as a result of regional manufacturing clusters, transoceanic freight, and the rapid global flows of objects as commodities and ultimately waste. These features of capitalist societies came about fairly recently through a process one of the authors termed the 'end of organized capitalism' in the mid-1980s in a book with sociologist Scott Lash. To some degree, this book also represents an attempt to reconcile the original thesis with the idea put forward by management theorist Gibson Burrell of 'reorganization' being a more appropriate idea than 'disorganization': 'It is one set of structures being transformed into another set of equally organized but different structures through a process of re-organizing'.[15] Here the term 'reconfiguring' is used in lieu of 'reorganizing'.

The rapid prototyping phase of 3D printing is an oft-unacknowledged enabler of what Nobel Prize-winner Joseph Stiglitz termed the 'roaring nineties' as the economy of the United States (US) shifted to services and manufacturing and was offshored within a free market system.[16] Without 3D printing in pre-production form, fit and function processes multinational corporations (MNCs) could not chase competitive advantages and leanness with the degree of profitmaking witnessed in late modernity, as the economies of scale in bulk volume production and distribution required the intensity of consumer demand to be matched by supply. The digitization of object designs and the spread of the Internet saw rapid prototyping as the solution of choice for MNCs to further de-concentrate their activities across great distances.

Most 3D printers on the market today use some sort of cartridge system in the same fashion as their 2D relatives. Just like in paper printers various manufacturers have sought to render their models compatible only with their own branded cartridges or those from preferred suppliers (many 3D printers also use standard powdered colour cartridges from 2D printer companies such as HP). As a rule the larger companies demand model-specific cartridge types and some even control the entire supply chain in-house.

3D printing design companies compete to offer user-friendly interfaces for customers with no experience in computer-aided design (CAD) software: these include Ponoko, Shapeways, Materialize and Fluid Form. But the real challenge is to facilitate the creative act of designing, including users in the design process, without users and co-producers learning complicated software packages such as 3D-CAD.[17] CAD skills are common for engineers and industrial designers. However, competency requires tuition and application. Efforts to make the graphical user interfaces (GUIs) of design software intuitive may bridge this hurdle. In recent years, the accessibility issues of design software have diminished. The application SketchUp is a 3D modelling system with a free version (Make) demanding no software coding and an online repository of designs. Other repositories of online designs exist including Thingiverse owned by Makerbot Industries (Stratasys).

Moreover, additive manufacturing offers designers the option of printing complex geometric designs that are more or less impossible in other forms of manufacturing where there are exotic materials that are difficult to use in a machining factory environment. Also it is possible to print organic matter such as food or organs for the body grown from a patient's cells.

Especially significant as already noted is that 3D printing enables the customizing of products. Unlike traditional manufacturing users can participate in the design of products. Even with increased printing speeds additive manufacturing will not be able to match the efficiency and speed of the global freight industry and next-day delivery models. But what it does offer is a way of producing bespoke and unique objects such as shoes tailored to the shape of each foot. This will set standards that cannot be matched by mass manufacturing. High quality low cost manufacturing systems may change how consumers shop and where objects are made, and this could cascade through the supply chain and logistics of freight companies.[18]

Moreover, what could emerge is downloading designs by users to replace broken or malfunctioning parts in a modular process where no technical tools or expertise are required. So in this future global manufacturing could be augmented or replaced by a circular economy where objects would be printed, used, and then locally recycled into further printable feedstock. Something like this has been imagined by the Ellen MacArthur Foundation, which flagged 3D printing as a key innovation in a future circular economy at the 2012 Davos conference of business elites.[19] With the development of infinite bandwidth and zero latency in online networks combined with personal fabrication the conventional trading pattern could be 'turned on its head', with artisans in the developing world 'crafting products for 3D printing' in the developed world and in the process re-engineering current craft value chains.[20]

Already there are prototypes of portable, laptop-style printers and these have the potential to be used on trains, planes, and on the go. It has become a cliché to highlight that the introduction of Internet, email, digital memory and computer word processing software and hardware was accompanied by great excitement about the redundancy and even obsolescence of paper, particularly within office settings. The 'paperless office' was foretold. Truly, paper remains an important part of many workplaces, however there has been a marked shift in its use. The increase in the sheer capacity of digital storage devices has meant paper records are rarely kept on site in archives for long periods. At the time of writing a gigabyte is the standard disk size for a ubiquitous USB 'thumb' drive – this is an impressive one thousand million bytes. The digitized reduction in physical space that digital storage affords means a reduction in the need for storage space, keeping in mind that most word processing or spread sheet documents, not including images or videos, remain under a megabyte (one million bytes) in size.

The transformation of office space is not simply the substitution of paper for digital information. While many workers continue to print out copies of online material for reading, the shrinking and even disappearance of physical archives – and the once-ubiquitous and now rare Manila paper folder – in many offices have been matched by the affordability of electronic paper printers, allowing digital documents to be printed in real-time. Alongside the massification of electronic printers are systems for paper recycling that coordinate with other waste disposal services. Image scanning is now a standard feature in electronic printers so that digital files for distribution via email or for digital storage are close to seamless. So rather than being paperless, offices are now 'printerful'. As children grow up with little use of paper and ubiquitous electronic printers in schools there could also be a surge in paperless spaces in the future.[21]

Objects made additively acquire their own identity from inception. There is then dissimilarity between objects of additive and subtractive origin – chiefly, the latter gain character through decay over time while the former are born with character. For instance, the scratches on a mobile phone are unique to that single object, similar to a fingerprint, particularly at a molecular level. Such identifying features in a 3D printed phone can be a part of the actual design. There is then an organic, evolutionary aspect to this process of additive manufacturing:

> Meanwhile, recent developments in rapid fabrication technologies (3D printing) and ever smaller and more powerful robotic platforms mean that evolutionary computing is now starting to make the next major transition to the automated creation of physical artefacts and "smart" objects.[22]

The additive process of 3D printing is also flexible, meaning materials are deposited in each layer with a concern for their surroundings. Many industrial manufacturing processes use heat to produce objects. Computer numerical control (CNC) machines use lasers to precisely cut away material until an object is made. Others use forges to heat solid materials until they are liquid and malleable enough to be beaten into different shapes.

So there are different species of 3D printing with each one entailing its own kind of innovation, technology, materials and processes. Each too will involve different social practices depending upon their affordances and technical features. Here we have used analogies to describe the nature of a few of these: SLA, fused deposition modelling (FDM), selective laser sintering (SLS) and electron beam melting (EBM). There are others in the experimental phase also, such as ionic covalent entanglement (ICE) gels that change over time once 3D printed in a predictable or controllable manner.

Not a single one of these innovations will be likely to reach ubiquity in their current manifestation. Instead, there will be combinations of these along with new affordances (for instance, multiple colour or multiple material printers) and the not yet imagined technical capabilities that are crucial in breakthroughs.

Yet what all of these developments have in common is the potential to reduce the division of labour, replace the need for assembly, render obsolete the manufacture of objects in bulk, and radically decouple transportation and energy from the production-distribution-consumption triad. These could have widespread social ramifications.

## System innovation

Thus there would be many significant social consequences if 3D printers were ubiquitous: it would represent a 'system innovation'. Personal fabricators potentially allow print-to-demand and even print-to-need social practices rather than conventional patterns of order, stock, supply and procure. In addition, currently popular practices in fashion, hobbies and craft of personalization, repair and customization may become mainstream within retail and leisure. Current efforts to upgrade Internet systems around the world towards near-infinite rates of bandwidth would impact markedly upon the viability of online systems enabling the sharing and sale of digital 3D designs by consumers themselves or by retailers and community resources who mediate on their behalf.[23]

There are important advances taking place in printers and the various components that are a part of them – guided extrusion nozzles and laser beams; motherboards, power supply units, memory chips and processors; and housing, frames and build trays. Materials are fed into printers in order to be made into objects and these can be powders, liquids, solids, and gases. Materials are derived from various resources including metals, petroleum, minerals, and even food.

Some future technical innovations are likely to include machines able to print mixed materials at the same time; the printing of active systems such as batteries, circuits, actuators and assembled machines; organic systems such as stem cells, organisms and cultures; infrastructures such as buildings, large structures and vehicles; and in situ objects inside the body, in space, in deep oceans, or whilst in motion. The range of materials that can be 3D printed is increasing rapidly.

Such 3D manufacturing has potential cost savings for business interests. These savings include: first, customizing objects for particular consumers; second, reductions in costs through 3D printing, or manufacturing, as demand dictates; third, being able to make small modifications to products at almost zero cost; fourth, savings in raw materials since little gets thrown away; and fifth, locally adapting designs to suit the parameters of particular environments. There is also potential for recycling both the excess materials and existing objects once formed. In 2015 Dutch start-up company Refil launched their fully recycled plastic filament wire made from shredded Volvo and Audi cars.[24] 'Fair trade' 3D printing filament is also a possible product from trials with waste-pickers in India.[25]

But the biggest saving is that consumers can manufacture on their own 'printer', or on one nearby, many objects they need or desire. This would go against the grain of the current ways government policy funds and supports manufacturing through big loans to major carmakers and other manufacturers. Instead what would be required are small injections of investment to support the quick manufacturing of complex objects through decentralized and distributed production technologies – what commentators term 'thinking small'.[26]

If small, dispersed centres of innovation reach fruition, 3D printing services on the high street or in shared facilities or possibly in the home could proliferate, all serving distinct functions and demands. Some would be niches and others would dominate. Overall there are many possibilities for a much greater localization of manufacturing: for some non-critical products the capacity to scan the object and then make endless copies, an 'infinite aisle', by or near consumers would produce large cost savings and decouple energy from freight, assuming that the same number of products was being manufactured worldwide and not drastically more.[27]

Potentially large savings in transport costs could mean that at some point low cost manufacturing centres, now for the most part sited in Asia, would no longer possess their comparative advantage in manufacturing. Digital objects can travel almost for free although oil is the basis for many of the powders used in such printing/manufacturing. Conversely there could be an intensification of transport as demand for raw materials for printing is added to the vast movements of manufactured objects by freight.

Thus a system innovation would be highly significant. Social thinker Frank Geels describes how major innovations in science and technology are often wide-ranging and not confined to the 'technical'. System innovations, such as the transformation of manufacturing, involve not just changes in technical products, but also he states: 'policy, user practices, infrastructure, industry structures and symbolic meaning etc.'.[28] This could lead to an 'after the factory' stage of development.[29] Web-based digital technologies could play a central role in global networks of digitally transferable and downloadable files containing designs and blueprints that home or office computers could then build, or print, anywhere and out of anything.[30] More critically, does 3D printing have the capacity to bring home some, at least, of the offshoring of work that has characterized the past thirty years?[31] Is additive manufacturing a world-changing innovation that would generate a new long wave of socio-technical change or is it simply a blip in the ongoing hype cycles around new technologies?

## The social

3D printing is in many ways a process of growth unlike casting, moulding and cutting, and offers new affordances for social individuation. Human-controlled growth, layer by layer over a time frame, excites the imagination because it is familiar in the complexity of the non-manufactured 'organic' landscape. Trees grow additively and no two are alike in an entire forest. Fungi grow additively and adapt to their unique surroundings. Insects, animals and indeed we humans grow additively and no two are alike, even in the case of 'identical' twins who share the same design through their DNA.

What these examples of 3D printing in organic, living entities show is that production in biophysical systems is radically different from bulk production in human systems. Industrial manufacturing produces many objects of identical form, that are then discarded, or recycled, once change sets in and compromises their integrity. This is not to say that objects in the past did not gain character through decay: there is huge market demand for 'aged', distressed or worn products ranging from wines to clothing. Yet a key characteristic of modern bulk industrial manufacturing is the prerequisite of cyclic disposal and replacement: so-called 'planned obsolescence'. This point is relevant because 3D printing is not just a new component of the current factory-centred system of manufacturing. In order to manufacture a single instance of an object's part, labourers must organize various subtractive 'tools' (milling, cutting, moulding machines) into unique configurations in industrial facilities at great cost. The expense and effort in so-called 'tooling' and 'retooling' are why objects are made in bulk volumes, and why batch quantities must decay quickly and be replaced regularly in order for profits to be yielded from what are considered by consumers to be 'disposable' objects. The short life cycles of many objects are coming under the scrutiny of both consumers and governments due to the commitment for corporations to be 'sustainable', although compliance and cooperation are difficult to enforce.[32]

With decentralized production in the form of 3D printing the user takes more responsibility for the choices made about what the final object will look like and involve, in terms of uses and social practices, before production occurs. Choices include layout, scale, cost per unit, quality, aesthetic detail, material composition, and so on. From a systemic viewpoint it can be expected that the current interlocking triad of production, distribution, and consumption systems will face transition. Various elements including, but not limited to, assembly lines, containers, transoceanic ships, inventories and malls will face upheaval as consumers manage their own printing processes on machines that are closer to them.

Another consideration is the labour and employment condition standards within the current system. On top of retooling costs in factories there is also the need for assembly once parts are produced in a subtractive way and in bulk. Profit is made through the sale of a great many identical parts, which can then be assembled by workers with little to no training or expertise: this is the 'division of labour'. Commodities are assembled at speed by lines of low-paid and untrained workers whose actions can be governed to microseconds. In recent times, the computer-control of both tooling and assembly, via robotic arms and other kinds of automation, intimated a system without human labour. However, this has so far remained a long way off because of the comparative advantage of manufacturing in Asia and other low-cost regions.

Hence there is another way of imagining manufacturing via 3D printers that require barely any assembly for those objects that can be printed by consumers themselves. These units would not function in large facilities but instead in a decentralized manner in order to make objects in single, or few, instances. The additive process of layer-by-layer growth is eminently suitable for objects that are geometrically complex, unique and printed to order, rather than stored within inventories and marketed to clear. Beyond supply chain 'leanness' such a vision is a radical overhaul of the current system of consumption, both in terms of profit-making and of imagining demand in consumer markets.

Sociologist Susanne Küchler was quick to realize how 3D printing results neither in a prototype, which can be serially reproduced in manufacture, nor a simulacrum of an object that exists already. Individuation of infinite variation poses a death-knell, one would imagine, for the current system, which thrives on obsolescence and would be turned upside down almost overnight.[33] Following Küchler, in this book the social is key to 3D printing. All forms of social life involve combinations of proximity and distance. These necessitate intersecting forms of physical, object, imaginative and virtual mobility that contingently and complexly link people in patterns of obligation, desire and commitment. These occur increasingly over geographical distances of great length.[34] Technologies too are rarely the result of single instances of inventiveness and inspiration. Instead they are a product of many combinations of innovators, experiments, and testers. Objects move to and fro between users and innovators as improvements are invented, faults revealed, fashions and trends wane, and decay results in an impact on functionality. In short, technologies are socially constructed and far from deterministic of societies and people.

Everyday life is increasingly becoming 'wrapped up' with digital information. This is happening in the form of small technologies worn on the body – intimately trusted by users – and those bolstering big systems that depend on them to function and 'synchronize'.[35] 3D printing appears to be a pending everyday technology and is thus an aspect of a wider social movement where digital information becomes materialized and empowers, or in some cases disempowers, users. Not all aspects of 3D printing are beneficial or a stage of progress. There are problematic consequences in the ubiquity of personal manufacturing technologies. 3D printers could be used by anyone, including children or teenagers, to print many artefacts currently licensed or prohibited, such as weapons or counterfeit goods.

There are abundant predictions heavy with hype: '3D printing may represent a disruption to the manufacturing industry as profound as the Industrial Revolution'.[36] For all of the 3D printing technologies available there are hurdles in producing large numbers of identical objects as economies of scale are difficult to achieve. 'Additive' forms of manufacturing are not a practical production alternative now or in the foreseeable future, compared to the speed and incumbent investment given to assembly line processes of injection moulding, metal casting or 'subtractive' CNC routing and milling. The outstanding hurdles for 3D printing generally are speed, accuracy, nonlinearity, material properties and system costs.[37] On the widely publicized consumer-level 'low-end' FDM 3D printers, there are issues with object strength, material mixtures, safety in home settings, and quality of appearance. For the industry-level 'high-end' SLS 3D printers, there are similar issues with the compliance and standardization of printed parts, limits to true multi-material processing, and the need for further stages in the integration of printed parts into final end-user products. However, despite these limitations significant potentials are emerging here.

## Why 3D print?

3D printers are an ever more common sight in engineering workshops, design studios, and even high street stores. 3D printers' possible affordances are beguiling; they are full of portentous features. Despite their allure many people face confusion in understanding their role in societies brimming with affordable, mass-produced objects. There is much commentary on 3D printing as a 'new industrial revolution' in mainstream media sources.[38] The excitement would suggest this technology offers an even faster, cost effective and bulk way to make large numbers of identical objects. In this book we argue that this is a misunderstanding of what 3D printing means for society and humans who might determine how this innovation be used. 3D printers are not just personal factories, simply replacing or substituting for the current methods, but also something else entirely.

An engineer asked the questions above on page 9 would most probably respond with another question. Possibly, but why would you want to make lots of identical items with a 3D printer? There are faster, more cost-effective and efficient ways for people to make objects en masse, she would respond. And in addition, due to limitations with 3D printing today there are many technicalities and challenges.

The first reason for hesitation is a material one. While plastic or resin 3D printers are now common and cheap, consumers are not currently able to purchase their own metal 3D printer in the same price range. Currently only industrial 3D printers are able to process metals. These technologies are expensive and so are the materials they use since they have to be processed industrially into powders or other forms of feedstock.

The second reason is to do with resolution. The majority of 3D printers are prone to some 'step' in the movement of the computer-controlled interface – that is, the extrusion nozzle or arm – so there is bound to be a degree of jaggedness on the edge. While some of the newer polymer methods give very high resolutions many metal objects require polishing or other kinds of finishing. Some might argue that resolution of the print is merely a cosmetic issue, however there are ramifications for the strength of the object and its ability to withstand wear over time.

3D printing is not just another stage in the evolution of industrial technologies, but also something else entirely. On analysis, it arises that at the industrial level 3D printing is or is not a viable alternative depending on the object to print and its applications. Although 3D printers offer the manufacturing of critical parts in metal the process would be far more expensive in comparison to current factory production and perhaps neither as strong nor resilient. There would need to be some very compelling reasons for consumers to use these tools instead of buying objects made in a factory from a shop. Some of these are reviewed in this book.

The issue of whether it is desirable to 3D print strikes at the heart of the commentary on its place in a new industrial revolution. 3D printers are just one of a number of fabrication technologies that manifest digital data in three-dimensions. The computer control of production machines considerably predates 3D printing and industrial ones – for instance, the straighteners, grinders and four-slides described in the next chapter – generally now incorporate digital circuitry of one sort or another into their designs. So are suggestions that 3D printers are a harbinger of a 'new industrial revolution' wrong or misguided? The answer here is no not necessarily. It is not impossible to imagine in future making common objects with near functionality to the originals on a 3D printer.

## 3D futures

In this book we present and assess four accounts of the future in 2050. In the next chapter we examine the potentialities of this system innovation, primarily in how the digitization of data is leading to decentralized forms of producing, distributing, and consuming objects. These four futures are developed from media commentaries, twitter feeds, science fiction, and insights from experts in the public domain presenting at conferences, interviews and opinion pieces. The scenarios were developed as self-contained 'worlds' determined along two axes: people's engagement with 3D printing and the degree of corporatization and openness.[39] In plotting these four worlds attention was paid to the systems and challenges emerging within societies up to 2050. The potential impacts of 3D printing for society could be significant, as inventor Sir James Dyson summarizes one kind of future:

> You can be independent. You don't need toolmakers. You don't need moulders. You don't need casters. You don't need foundries. You can do it all yourself with a relatively simple (I hope) machine. So you can make things all over the place. You can make them very locally to each country that you are selling it and get rid of freight costs and import duties and all those sort of things. I think it will eventually transform the ways products are made.[40]

Thus this book tries to show how 3D printing could be more than a notch in the evolution of technology, that it has many societal features with capacities for major change. One future would involve the end of trucks and freight trains, while container ports stand empty. Containers would rest rusty and hollow on the dock. The assembly lines would be quiet and the warehouses still. People's homes would now be filled with activity as they download and 'print' the objects they desire like pieces of paper on a home printer. And once people were done with these objects they would recycle them, reusing the same materials again and again. And most designs and printers would be open source and available online. Some people would just swap and share designs, others would contribute their own expertise to communities of fellow innovators and databases of shareable designs, ranging from a new kitchen sink plug to a replica antique train model.

Now imagine another future – one filled with more local deliveries and freight than ever before, influenced by the return of regional industry and the ubiquitous spread of the digital 'factory in a box' to retailers. People would print out objects at specialist shops, which would rely on standardized and regulated supply chains delivering all sorts of industrial materials, from metal powders and polymer filaments to exotic liquids and resins. The ease with which objects could be printed would lead to an intensification of movement, with roads and railways busy distributing the many items people constantly order online through a plethora of corporate suppliers and finance systems. This is a world where objects are cheap, materials are moved about rapidly, and products quickly become obsolete.

In these two polar futures a significant social and technical transition has taken place in the production, distribution and consumption of material objects. Regardless of which future is more 'real', they both have in common an innovation allowing virtually transmitted digital bits of information to be made into real-world atoms. These are just two futures of a possible many representing the conjunction of digital information, automation, materials science, and computer-aided design through 3D printing or more broadly 'layer-by-layer' additive manufacturing. This process for making things represents a possible socio-technical transition in manufacturing and transportation, not to mention marketing. Turning bits into atoms in real-time is as profoundly unfamiliar as the idea of instantly sharing thoughts across the world would appear to a nineteenth-century observer who would marvel at this readily available feature of modern communications.

## Structure of the book

In this chapter we have considered various affordances and limitations of 3D printing. More than just another manufacturing technology on the assembly line we argue in this book that a new system is coming into being. What this new system will be like will be determined by the nature of this engagement of social actors with 3D printing and the degree of openness of the new technology and the data it involves. This book thus explores a system 'innovation' or indeed 'reconfiguration' in response to uncertainties facing societies both now and in the near future.

3D printing will not determine society in the future. If a three-dimensional society is to emerge wherein this particular innovation profoundly disrupts or reconfigures the current triad of production, distribution and consumption, then it is crucial to understand its potentialities and the nature of such a decentralized system. Such an outcome will be the goal of the next chapter where the potential for 3D printing to become ubiquitous is assessed through an examination of its history. Throughout the twentieth century various predictions arose about the effects of automating production upon different aspects of society: employment, mobility, equality, and so on. Changes to dominant social norms were also articulated as being under pressure, for instance what would people do with no work and much more leisure-time? Chapter 3 considers how 3D printing is emerging within the current triad of production, distribution and consumption through the compression of space and time, concerns about offshoring and uncertainties from supply chains and global production networks and metamorphosing systems of online and offline retail. In Chapter 4 the issue of resources is examined through analysing the materials that 3D printers depend upon. The rhetoric around 3D printing's disruptive potential does not give due consideration to sustainability and the environment in terms of waste and resource extraction. Chapter 5's focus is upon social transformation and the possible socio-technical transition of which 3D printing is a key part in many future projections. Chapter 6 builds on this by detailing a futures scenario workshop and qualitative interviews with experts. The final chapter fleshes out the different scenarios drawn from this research based on two axes. The chapter tries to assess just what kind of niche 3D printing occupies, and whether it will result in regime change and the transformation of the entire landscape of manufacturing, transport and work.

So to conclude, 3D printing could reconfigure the current complex through drawing consumers away from the kinds of products currently supplied to them and unsettling suppliers to the point of inciting new business models and ways of making profit.[41] The disruption would have knock-on effects in user understandings of the value of materials (likely to be in the form of cartridges of powders) and energies, and how objects become recycled or simply waste for landfill once their life is over. One significant feature of 3D printing is that it allows people to manipulate matter in a new way. It allows the transfer of ideas between a computational, virtual space and the material world. This gives people the ability to use more complex processes to design, share and model objects, and then give them form in almost any material. This opens up a range of possibility in what can be produced – people can create objects that, previously, could not have been produced, or even imagined. 3D printing both allows and invites people to extend their imagination as to the types of things that can be produced.

In the next chapter we consider the ubiquity of 3D printing and the shifts in labour and society more broadly that could result. We will briefly consider how 3D printing emerged from within the current system of production, distribution and consumption in the 1980s and 1990s. 3D printers are not fictional technologies; they are currently used in the manufacture of nearly all of the products on the global mass market. 3D printing has emerged from within the current system for producing objects in the general trend towards the automation of pre-production. The reason for this origin is the on-going demand for 'leanness' and 'flexibility' that results in efforts to compress time and space. In the 1980s, within sections of industry, there arose a window of opportunity for prototype testing that departed from either craft expertise – that is, hand modelling in clay, wood or stone – or digital CAD file creation and factory production, including costly factory re-tooling, time-consuming transoceanic freight, and patent protection for intellectual property released before the product went to market, thereby increasing the risk of forgeries and 'knock-offs'. Both of these options, namely for the pre-production of models, added onerous uncertainties, delays and costs.

## Notes

1 N. Savage, 'Technology: Building Opportunities'. *Nature* 509 (2014): 521–23. T. Wohlers, 'Tracking Global Growth in Industrial-Scale Additive Manufacturing'. *3D Printing and Additive Manufacturing* 1, no. 1 (2014): 2–3, doi:10.1089/3dp.2013.0004

2 K. Tucker, D. Tucker, J. Eastham, E. Gibson, S. Varma and T. Daim, 'Network Based Technology Roadmapping for Future Markets: Case of 3D Printing'. *Technology and Investment* 5, no. 3 (2014): 137–56, doi:10.4236/ti.2014.53014

3 M. Raby, '3D Printing Market to Hit $3 Billion by 2018', 2012 New York. Accessed 3 August 2012. http://www.slashgear.com/3d-printing-market-to-hit-3-billion-by-2018-23239870/

4 J. Rifkin, *The Age of Access: How the Shift from Ownership to Access is Transforming Modern Life*. London: Penguin, 2001.

5 C. Hull, 'Stereolithography: Plastic Prototypes from Cad Data without Tooling'. *Modern Casting* 78, no. 8 (1988): 38.

6 S. Mellor, L. Hao and D. Zhang, 'Additive Manufacturing: A Framework for Implementation'. *The Economics of Industrial Production* 149 (2014): 194–201, doi:10.1016/j.ijpe.2013.07.008

7 Makerarm, 'Makerarm – the First Robotic Arm That Makes Anything, Anywhere', 2015 Kickstarter. Accessed 12 October 2015. https://www.kickstarter.com/projects/1849283018/makerarm-the-first-robotic-arm-that-makes-anything

8 K.V. Wong and A. Hernandez, 'A Review of Additive Manufacturing'. *ISRN Mechanical Engineering* 2012 (2012): 10, doi:10.5402/2012/208760

9 MetalBot, 'The Challenge … To Make the World's First Open Source Metal Printer!', 2015 Accessed 30 October 2015. http://www.metalbot.org/

10 K. Tucker, D. Tucker, J. Eastham, E. Gibson, S. Varma and T. Daim, 'Network Based Technology Roadmapping for Future Markets: Case of 3D Printing'. *Technology and Investment* 5, no. 3 (2014): 137–56, doi:10.4236/ti.2014.53014 139

11 B. Mosadegh, G. Xiong, S. Dunham and J. Min, 'Current Progress in 3D Printing for Cardiovascular Tissue Engineering'. *Biomedical Materials* 10, no. 3 (2015): 034002.

12 Q. Zhang, X. Yang, P. Li, G. Huang, S. Feng, C. Shen, B. Han, X. Zhang, F. Jin, F. Xu and T.J.Lu, 'Bioinspired Engineering of Honeycomb Structure – Using Nature to Inspire Human Innovation'. *Progress in Materials Science* 74 (2015): 332–400, doi:10.1016/j.pmatsci.2015.05.001

13 W.B. Arthur, *The Nature of Technology: What It is and How It Evolves*. New York: Free Press, 2009.

14 F. Jia, X. Wang, N. Mustafee and L. Hao, 'Investigating the Feasibility of Supply Chain-Centric Business Models in 3D Chocolate Printing: A Simulation Study'. *Technological Forecasting and Social Change* 102 (2016): 202–13, doi:10.1016/j.techfore.2015.07.026

15 G. Burrell, 'Book Review Symposium: Scott Lash and John Urry the End of Organised Capitalism'. *Work, Employment & Society* 27, no. 3 (2013): 537–38, doi: 10.1177/0950017013479554. 538

16 J. Stiglitz, *The Roaring Nineties: A New History of the World's Most Prosperous Decade*. New York: Norton, 2003.

17 A. Chen, '3-D Printers Spread from Engineering Departments to Designs across Disciplines'. *The Chronical of Higher Education*, 2012. Accessed 17 September 2015 http://chronicle.com/article/3-D-Printers-Arent-Just-for/134440/?cid=at&utm_source=at&utm_medium=en

18 D.W. Rosen, 'Advanced Technology and the Future of US Manufacturing', Atlanta, 2004.

19  E. MacArthur, 'Towards the Circular Economy'. 2012 McKinsey and Company. Accessed 15 August 2012. http://www.thecirculareconomy.org

20  S. Bell and S. Walker, 'Futurescaping Infinite Bandwidth, Zero Latency'. *Futures* 43, no. 5 (2011): 525–39, doi:10.1016/j.futures.2011.01.011. 532

21  J. Pedersen, 'Project Work in the Paperless School: A Case Study in a Swedish Upper Secondary Class'. *Education and Information Technologies* 9, no. 4 (2004): 333–43, doi:10.1023/B:EAIT.0000045291.99489.bd

22  A.E. Eiben and J. Smith, 'From Evolutionary Computation to the Evolution of Things'. *Nature* 521, no. 7553 (2015): 476–82, doi:10.1038/nature14544. 480

23  S. Bell and S. Walker, 'Futurescaping Infinite Bandwidth, Zero Latency'. *Futures* 43, no. 5 (2011): 525–39, doi:10.1016/j.futures.2011.01.011. 532

24  Refil, 'Refilament', 2015 Better Future Factury – Perpetual Plastic Project. Accessed 21 September 2015. www.re-filament.com.

25  T. Birtchnell and W. Hoyle, *3D Printing for Development in the Global South: The 3D4D Challenge*. Basingstoke: Palgrave Macmillan, 2014.

26  V. Wadhwa, 'How to Save the Global Economy: Think Small', by Vivek Wadhwa, *Foreign Policy*. Washington, DC: FP Group: The Washington Post Company, 2012. 3 January 2012. Accessed 3 October 2015. http://www.foreignpolicy.com/articles/2012/01/03/8_think_small

27  C. Anderson, *The Long Tail: How Endless Choice is Creating Unlimited Demand*. London: Random House Business Books, 2012. 226.

28  F.W. Geels, 'Multi-Level Perspective on System Innovation: Relevance for Industrial Transformation'. In *Understanding Industrial Transformation: Views from Different Disciplines*, edited by Olsthoorn and Wieczorek, 163–86. Dordrecht: Springer, 2006. 165.

29  S. Fox, 'After the Factory [Post-Industrial Nations]'. *Engineering & Technology* 5, no. 8 (2010): 59–9, doi:10.1049/et.2010.0814

30  H. Lipson, 'This Will Change Everything'. *New Scientist*: New Scientist, 2011. Accessed 3 March 2012. http://www.newscientist.com/issue/2823

31  G.F. Davis, *Re-Imagining the Corporation, Real Utopias*. Ann Arbor: The University of Michigan, 2012.

32  S. Freidberg, 'It's Complicated: Corporate Sustainability and the Uneasiness of Life Cycle Assessment'. *Science as Culture* 24, no. 2 (2014): 157–82, doi:10.1080/09505431. 2014.942622

33  S. Küchler, 'Technological Materiality: Beyond the Dualist Paradigm'. *Theory, Culture and Society* 25, no. 1 (2008): 101–20, 58–9, doi:10.1177/0263276407085159. 109

34  J. Urry, 'Mobility and Proximity'. *Sociology* 36, no. 2 (2002): 255–74, doi:10.1177/0038038502036002002. 256

35  T. Birtchnell and J. Urry, 'Small Technologies and Big Systems'. In *The Mobilities Paradigm: Discourses and Ideologies*, edited by Endres, Manderscheid and Mincke. Farnham: Ashgate, 2016.

36  I.J. Petrick and T.W. Simpson, '3D Printing Disrupts Manufacturing'. *Research Technology Management* 56, no. 6 (2013): 12–16. 12.

37  J. Bhattacharjya, S. Tripathi, A. Taylor, M. Taylor and D. Walters, 'Additive Manufacturing: Current Status and Future Prospects'. In *Collaborative Systems for Smart Networked Environments*, edited by Camarinha-Matos and Afsarmanesh, 365–72: Berlin/Heidelberg: Springer, 2014. 366.

38  P. Markillie, 'A Third Industrial Revolution'. 2012 The Economist Newspaper Limited. Accessed 14 September 2013. http://www.economist.com/node/21552901

39  M. Linz, 'Scenarios for the Aviation Industry: A Delphi-Based Analysis for 2025'. *Journal of Air Transport Management* 22 (2012): 28–35, doi:10.1016/j.jairtraman.2012.01.006

40  R. Cellan-Jones, '3D Printing – a New Industrial Revolution?'. 2012 BBC. Accessed 30 October 2012. http://www.bbc.co.uk/news/technology-20137791

41  I.J. Petrick and T.W. Simpson, '3D Printing Disrupts Manufacturing'. *Research Technology Management* 56, no. 6 (2013): 12–16.

# 2

# A BRIEF HISTORY OF 3D PRINTING

## Introduction

A 1935 newsreel from British Pathé foresaw a future when the city composed of 'a vast agglomeration of tall buildings and slums' is radically transformed. It would be architecturally planned to a new order and less prone to 'pure and rugged individualism'. 'Cities of the future would be laid out to a master plan', one which contained 'artistic streets: there must be broad avenues providing trees and ease of movement with light and pure air for all buildings' the newscaster muses. The 'master stroke' in this is the 'highly efficient automatic industrial centres producing all our needed goods and distributing them in abundance', segregated entirely from the 'new centres of leisure' where people live.[1]

Such an idea built upon the future projections made a decade earlier by visionary 'starchitect' Henry W. Corbett in his 'Wonder City You May Live to See'. He foresaw congestion in megacities being solved by utilizing all three dimensions of urban space for transportation. The street level is reserved for pedestrians while slow and fast private and public vehicles travel along underground tunnels or move between rooftop landing platforms (for zeppelins). But the most beguiling feature of his future is that on the lowest level there are 'freight tubes' for automatically transporting parcels and goods.[2]

While neither pipelines for objects nor automatic industrial centres at the heart of the city have quite developed, there is something similar now developing with citizens procuring products with little human intervention. The system innovation contemplated in this book is foreshadowed by the seemingly magical possibility of objects becoming digitized similar to music, text and video. The physical appearance of an object in real-time (although not always rapidly and often somewhat slowly) from a two-dimensional virtual computer design is the result of this digitization of 3D data.[3] As surmised in Chapter 1 the use of the term 'printing' is not trivial. Since liquids were the initial material these technologies used in SLA – that is, in 'photo-polymerization' – there was an obvious similarity between how paper printers laid ink onto a page, as with 3D printing the repeatedly 'stacked' layers eventually create an object once the liquid is 'set' through laser curing.

This chapter examines the '3D printing system' that is coming into being around the world. Just like British Pathé and Henry W. Corbett we are concerned here with the future. And this is a future which could have profound consequences for global patterns of manufacturing, transportation, distribution and consumption. And it raises very significant issues of anticipating a future that is already in the process of formation. Just how significant and transformative could this system change turn out to be? We examine how 3D printing has already altered society and made the triad of production, distribution and consumption leaner and to some extent automated.

In this chapter we assess the ease with which rapid prototyping slotted into the triad of interlocking systems providing the lion's share of the world's objects. With the arrival of information and communication technologies in the current production–distribution–consumption triad came new possibilities for industries to experiment with manufacturing processes from a computer and digital fabrication technology rather than in a factory. Rapid prototyping (pre-production 3D printing) became an efficient solution to having to send objects not yet ready for market via long-distance freight, underpinning the centralization of manufacturing in key regional clusters. 3D printing is not so easily going to become a feature of the current triad since it intimates greater decentralization.

The history of manufacturing is a shift from decentralization to centralization. Here access to the means of production through cottage industries, skilled artisanal labour and traditional communal factories receded in favour of large-scale, automated 'assembly line' processes, with changes in machine tools, power generation, materials handling technologies, building materials, and power-distribution technologies.[4]

## Not so trifling manufactures

The manufacture of pins has been a lodestone for thinkers on progress in industrial production for centuries ever since inspiring the prominent economist Adam Smith. In the eighteenth century the 18 (or so) distinct stages in pin production utilizing the latest machinery propelled Smith to offer commentary on the impact of organizational and technological ingenuity working in tandem. As the quotation below illustrates, Smith was obviously impressed by the affordances made available by the division and mechanization of labour:

> To take an example, therefore, from a very trifling manufacture; but one in which the division of labour has been very often taken notice of, the trade of the pin-maker; a workman not educated to this business (which the division of labour has rendered a distinct trade), nor acquainted with the use of the machinery employed in it (to the invention of which the same division of

labour has probably given occasion), could scarce, perhaps, with his utmost industry, make one pin in a day, and certainly could not make twenty.[5]

The efficiency and mechanization of industrial pin making gave Smith food for thought. Some of the stages he noted were how the metal wire was drawn out by hand, then straightened and cut, before being sharpened on one end and then ground down. The head was fastened on and then the whole pin was coated to make it resistant to corrosion. Groups of pin-makers, each working on distinct manufacturing stages with specialist technologies, made many more pins much faster than a single pin-maker using craft expertise alone. Cost, speed and volume were benefits. Smith surmised that the combination of mechanization with organization was ushering in a new era of material wealth.

Proving him right, industrial pin making went up a gear a century later as did the manufacture of a great many other objects in this manner. The Industrial Revolution in the nineteenth century saw large factories emerge for the greater division of labour and much larger machines with many more stages alongside them. The use of steam and eventually electricity from fossil fuels took production beyond the limits of human, animal or micro-power generation – that is, water- or wind-mills – providing further cost-effectiveness, volume and speed as these machines neither slept nor got tired.

Nowadays automatic machines remove the need for most human labour altogether in pin making. Specially tooled factories utilize computer-controlled milling machines and automatic assembly lines, chemical polishers and industrial sharpeners to make pins more cheaply, faster, and in more volume than ever before. In comparison to the 18 stages of the eighteenth century usually four machines are required now to make pins in the twenty-first century. The process generally involves a roll straightener machine that cuts steel wire rolls into pin lengths. Conveyor belts then take the cut wire to a second grinder machine for polishing and sharpening. A third four-slide machine stamps heads from sheet metal. Finally, a machine with a rotating tank chrome plates the pins and coats them with chemicals to provide a finish. Sometimes pins are sorted by human hand for quality assurance in order to spot defects – a lonely job indeed for a person in the company of machines.

The latest manufacturing technology, the subject of this book, intimates an entirely different way of making pins: additively on just one machine. Instead of cutting pins from wire (subtracting them from the base material) they could instead be built up in three dimensions layer by layer without any assembly at all and in situ. 3D printing involves fusing deposits of powder using a computer-controlled laser, much like the laser printers used for text and image production that are common fixtures in offices and homes today. So could pins *actually* be made with a 3D printer by current standards? Our argument in this book is that there are powerful indicators in the current system that key uncertainties in the twenty-first century will be met by a reconfiguration at the scale of societies involving 3D printing. But first, in this chapter we establish what 3D printing is and summarize the things this innovation can and cannot do by looking more closely at its history.

The move of fabrication from individuals and communities in the Global North to centralized industrial settings under the control of a few elites in the first Industrial Revolution has been well-documented elsewhere and lies beyond the scope of this book; however pertinent here is an analysis on the long tail socio-technical transition to 'Fordist' factory production in the US and other regions in the Global North throughout the early to mid-twentieth century.[6] We take up the story again in the mid-twentieth century where production is being automated and there is a widespread concern about the disappearance of manufacturing work due to new technologies.

## Automation

A key feature of the history of 3D printing is its role in automating human labour, notably the hand modelling involved in pre-production in the 1980s, but also before that with the shift from humans to machines in factories. Since the mid-twentieth century the concern over whether automation technologies will substitute for human labour has grown. 3D printing certainly automates many processes originally done by hand, yet rather than replacing people 3D printers also augment and even empower them. How can we understand this feature of the innovation? In considering apprehension about the scope of automation to substitute for prevailing forms of human interaction with manufacturing, we must inquire: do technologies ever entirely substitute for anything? Questions like this already have their own legacy in the social sciences, predominantly in rejections of technological determinism. Technological determinists argue that new technical developments determine societies: the evolutionary sequence of the Stone Age, Bronze Age and Iron Age is a classic example of this way of thinking. More seriously, determinists suggest that social, economic, and environmental issues are solvable through technological solutions. Global warming is fixable through 'geo-engineering' the climate;[7] transport congestion is fixable through 'intelligent' traffic modelling software.[8] Both of these options are patently unrealistic in practice, however convictions of the potential of automation to techno-fix the world are rife in global, digitized neoliberal societies.[9] In contrast, many social scientists argue that technologies are socially constructed and have different effects across space and time; social problems cannot simply be solved by technology but also require social and political solutions.[10]

There are numerous examples of technologies that are socially constructed: bicycles, QWERTY keyboards, automobiles, and so on. We have our own favourites here too. The services of a dental filling or facing are certainly comparative to those that teeth provide for eating or experiencing taste. Artificial turf is a common substitute for grass in sports venues due to lower costs and greater resilience. Electronic synthesizers and drum machines surely do replace drummers and other band members in genres of electronic music. As these examples show, while substitutive technologies do impact upon human actions, there is a two-way relationship, as humans and societies also impact upon technologies' designs, uses and contexts. So a middle position is to acknowledge that substitution does occur – however, this is not to say that these technologies are of the same nature, kind, or quality as those phenomena that they replace. While fillings give the added benefit of alleviating discomfort from caries, few are likely to trade in their teeth from birth for a set of dentures. The *Fédération Internationale de Football Association* (FIFA) continues to elect grass for their World Cup fixtures – the 2015 Women's World Cup was a provocative exception – even for 2022's water-scarce host Qatar.[11] The composers of popular genres of electronic dance music continue to reuse and repurpose sound samples – Gregory Coleman's 'amen' is a popular choice – of drummers from the 1960s due to their unique character and feel that are impossible to replicate.[12] All of these examples show that socio-technical transitions do not turn out to be a like-for-like substitution.

Even if such a response would alarm those people who depend every day on technologies that replace human actions, the concern is not sufficient for them to rush to abandon the affordances that technologies do provide. Technology does not determine human actions; humans determine the applications of technology. Social and cultural forces are just as important in the development of technology as economic or technical ones. Keeping these points in mind, it is fruitful to understand technologies such as 3D printing not only as potentially substitutive, but also as enabling or becoming a part of different socio-technical systems.

The ubiquity of automation in financial transactions demonstrates forcefully that technologies are socially constructed. Automation technologies that substitute for physical traders in stock markets on the trading floors do not necessarily replace human labour, instead they alter the time and organization of the entire trade system in all sorts of ways that were originally unforeseen. Sociologists Donald MacKenzie and Juan Pablo Pardo-Guerra develop this idea in their micro-history of the Island: a mediocre automatic trading innovation initially introduced to compensate for – and capitalize on – glitches in the US 'National Association of Securities Dealers Automated Quotations' (NASDAQ) with trading eighths of dollars.[13] The Island set a standard now ubiquitous across the world, notably in the integration of electronic trading innovations into the trading system itself.

There is a second common theme to examine in considering the scope for 3D printing to impact upon societies in the future. Industrial technologies homogenize people, place and products: 'as the assembly line produced identical goods, it seemed to erase difference'.[14] Sociologist David Nye gives another possible reading of the impact of industrial manufacturing. When we consider the future of 3D printing in terms of demassification, we must be mindful of the system of massification that brings 3D printers to market and allows consumers and professionals to utilize them. The homogenization thesis (mass manufacturing means less choice, culture and variety) is only half the story, and in a similar fashion to the substitution thesis there are many complexities in how people use technologies and engaged with industrial manufacturing in the past. 3D printing's ubiquity is founded upon the notion of heterogeneity: that the process will allow consumers to decouple themselves from the mass manufacturing system in favour of bespoke, personalized, 'mass customized' object production technologies. How are we to understand this idea of a retreat from homogeneity if this is a too simplistic history of mass manufacturing technology?

## Substitution

The notion that a substantial number of jobs are imminently about to be made redundant by automation is an enduring theme of recent history, at least in the Global North. Predictions of impending employment catastrophe are difficult to reconcile with current experiences. Work for wages continues to preoccupy more than ever regardless of whether technologies allow for more time through their greater efficiencies. It appears that new jobs have appeared to replace those that were automated. A cursory survey of the job market shows that occupations are far from scarce, if more precarious. Automation surely has taken place, however not with the kinds of consequences foreseen five decades ago. Moreover, what late nineteenth-century sociologist Max Weber termed the 'Protestant Ethic' is as strong as ever, namely the spiritual 'guilt' many experience when not undertaking routine tasks on a daily basis.[15] If 3D printing is a form of automation how does it fit in with these forecasts?

According to industry commentators, the reason why automation did not diminish employment is that the affordances and limitations of the specific technologies were misunderstood and, moreover, the ways they were socially constructed rather than deterministic of change were misconstrued. As *The Economist* summarizes, manufacturing – to take one sector – still needs people with specialist knowledge to service machines and set them up. Furthermore, assembly line robots continue to remain too dear for small and medium-sized businesses across the world.[16] As well, creativity and very many cultural pursuits continue to hold currency, particularly those that create new and unforeseen innovations. So there are a number of sociological explanations for why automation has not eroded employment as anticipated.

In 1964 it was hard to imagine what an automated society might be like to live in. Industrial computers – able to replace human 'brain' work to varying degrees of success – were only available to a small number of scientists and other technical specialists. Personal computers were not yet a ubiquitous consumer item. Those machines able to replace human 'muscle' work were more visible in the public domain in the form of automated elevators, white-goods, factory equipment, and so on. A popular idea in the social sciences and philosophy at this time was that as machines became more complicated and 'smarter' they would begin to replace brainwork as well as muscle work. Yet, it was uncertain how this would affect society beyond anthropomorphism: machines would become more human.

The book *Technocracy* by Jean Meynaud, published in 1964 and translated from French in 1968, demonstrates the understanding at this time of the cutting edge of automation. The automation of brain work is referred to briefly as synonymous with the emergence of electronic computers: 'Naturally the largest number of computers is to be found in the United States, where the number has increased from 100 in 1951 to 22,500 today'.[17] There is precious little extra detail about the computer's scope to replace human intellectual tasks in an industrial or office context, evidently because of this scarcity. Did Meynaud ever encounter a computer himself while writing his book? Obviously automation was understood more ideologically that technically, that is, as an ominous threat to individual liberty and human autonomy.

Many of the early commentators on the impact of automation on society were the first cabs off the rank and this shaped their ideas. One of the most important books in this genre was Herbert Marcuse's now classic *One-Dimensional Man*. There are many cardinal ideas within this work, principally the notion that massification extends beyond simply production, distribution and consumption. We examine Marcuse's ideas and then historicize these in relation to forecasting about automation in the mid-twentieth century and its anticipated impacts on manual labour. We then look at the turn to the offshoring of work and its relationship to automation, which we argue disrupted the expected trajectory. We then go on to consider the introduction of 3D printing as 'rapid prototyping' and its growing potential for 'demassification', and conclude with a summary of the main themes of the chapter.

In hindsight, the fear that brainwork is soon to be upstaged by computer automation is more than ever in the public glare. A feature piece in *The Economist* on the future of jobs with the title 'The Onrushing Wave' describes a new era of automation enabled by ever more powerful and capable computers, triggering a surge of so-called 'technological unemployment' in its wake.[18] It appears that the foresight of the 1960s was cognizant of the limits and horizons of technology.

Up until the mid-twentieth century the objects people used in the Global North were still made by people who operated large machinery, typically in workshops or factories. By the 1960s computer control in factories resulted in automated mass manufacturing. Manufacturers were incentivized to standardize and mechanize assembly lines. During the golden age of Detroit, the car manufacturing capital and model of computerized factory automation, large-scale manufacturing controlled by computers and involving on-site production and assembly took place thorough increasingly large corporations.

Even though the mechanization of labour has been a subject of speculation at least since the 1950s, it was not until the 1960s that automation seemed to be on the brink of transforming societies. US President Eisenhower famously quipped, 'Whatever saves the time and the effort of humans does give them greater opportunity for self development in their moral, their intellectual, and cultural sides'.[19] An animation from the feature mentioned above in *Life Magazine* from 1959 shows labour being pushed by automation into non-automated factories and to services. Services, and 'unproductive labour' more generally, did indeed replace the declining profitability of the manufacturing industry as predicted in 1959 in the now EU and US throughout the 1980s in 'post-industrial' regions.[20] Yet the impacts of automation were thought to be more goods made available to consumers, unemployment for a 'few' factory workers, more profit for manufacturers, and more leisure time and 'the good life' for all. The latter is certainly not the case. So what happened instead?

Differences in labour conditions and incomes across the world disrupted this idea of a rapid shift from people to machines. Since the 1960s the shift of production to countries in the Global South has led to a slow transition to automation, as factories in the Global North are unable to compete with off-shored manufacturing involving a vast supply of labour in Mexico, Panama, China, India, Vietnam, and elsewhere.[21] So a very different scenario emerges wherein instead of machines replacing manual labour entirely, manufacturing labour is offshored to countries where manual labour is more profitable than automation in the Global North. Cost savings are the main factor that drives offshoring and it is through non-unionized labour, less regulation and lower tax rates that profits are made.[22] Free trade zones within these nations became a method for states with little to no geopolitical allegiances to cash-in on off-shoring. Since the 1990s a further progression is the offshoring, rather than the automation, of white-collar knowledge work to countries where 'comparative advantages' offer profit.

In both these cases of the offshoring of blue and white collar work, technological development, including digital information and communication technologies (ICTs) that allow the close tracking, auditing and surveillance of workers across the globe, has given vast numbers of people access to labour opportunities – exploita-tive or otherwise – in the Global South, while at the same time destabilizing tradi-tional manufacturing regions. Computer control has found an application in the regulation and efficiency of manual, human workforces, rather than in their substi-tution. Technological innovations in manufacturing have occurred alongside the science of logistics and the containerization system, which has reshaped production and distribution with war-like efficiency.[23]

Marcuse, similar to many of his contemporaries – including Daniel Bell and Alvin Toffler – found intellectual energy in forecasting a 'post-industrial society' where people would no longer have much direct contact with production other than being compelled to purchase ever more automation-made things. The auto-mation of both manual and intellectual work would lead to a one-dimensional society shaped by its access to anything the imagination could muster in a sort of systemic feedback loop. A technological a priori societal ensemble circumscribes culture in its entirety as human-made creations are consumed and foster ever more desire for consumption at the expense of individual autonomy, or so Marcuse's thinking goes.

One of the most profound underlying assumptions of Marcuse and many other commentators in the mid-twentieth century was that machine labour would sub-stitute for human labour in manufacturing and this notion guided many of their key ideas. It is clear that this was a far too simplistic view of industrial automation. While certainly human labour in manufacturing did diminish in the Global North in what are now 'post-industrial' or 'deindustrialized' societies, this was a consequence of offshoring, rather than substitution, as human labour increased in manufacturing to an extraordinary degree in Asia and other manufacturing regions hosting compara-tive advantages for multinational corporations (MNCs).

Two key questions for twenty-first century production are: will automation substitute for the vast cohorts of human muscle-power that are in employment in factories in manufacturing regions in today's world where production was offshored to? Furthermore, what will be the spatial and social consequences of the automation of muscle-power if this indeed does take place? Commentators in the mid-twentieth century on automation understood there to be a trajectory from the substitution of muscle-power by machines to the eventual substitution of brain-power as well once technologies attained a level of sophistication.

First, it needs to be emphasized that a number of trends suggest that centralized automation could shift to decentralized automation. With the massification of global communication technologies (principally the Internet and smart phones) people are able to both create and share content with ease. A generation of 'smart' materials, including liquid metals, polymers and nano-fibres, are now entering the consumer market, with ramifications for object customization. The 'shrinking' of technologies, made both more powerful and portable, would allow consumers to own or access factories-in-a-box. Also, the standardization and aggregation of many digital object design 'files', available both freely and at cost, diminish the need for brand ownership and the production of many items of identical form.

Second, the machine automation of muscle and brain labour does not necessarily substitute for humans resulting in a reduction of work for them. Rather it simply changes the nature of work and in some cases intensifies it. Take, for instance, labourers in China working on assembly lines in factories. They are expected to perform on a par with machines, interfacing with various technologies that assist them to work more efficiently, managing their time, monitoring their activities, and regulating their physical movements down to the second and even the millisecond. It is far more profitable in the present day to employ cohorts of these interfaced, automated humans than robotic 'workers'. Similarly, with brainpower, automation of intellectual functions has not reduced workloads. Take the case of referencing software that automates bibliographies. All this does is make time for more content to be created and greater volumes to digest. In the same fashion as factory labourers, the greater surveillance of knowledge workers is another by-product of automation – for instance, metrics databases that evaluate academic performance based on the quantity and quality of outputs.

Studies of the impacts of teleconferencing and telecommunications technologies show these do not simply substitute for transport demand.[24] The same conclusions are applicable to 3D printing, which will not simply substitute for transportation in the current system of production, distribution and consumption: it could 'decouple' transport from object procurement.[25] Chiefly, the disruption will be through the decentralization of the means of production to consumers either in their homes or close by in dedicated facilities and service providers – 'bureaus', in current terminology.[26] Certainly, there will be geographical and indeed geopolitical implications as global supply chains and production networks face competition from decentralized personal production.[27]

Indeed we learn from studies of the impact of tele- and video-conferencing on transportation, long thought to be an obvious replacement for physical commuting, that substitution is not always straightforward and that new technologies can increase rather than decrease demand.[28] For instance, mobile technologies give commuters the capability to use a larger variety of spaces, often in real-time; offer more efficiency in their travel time; and provide information, entertainment and multitasking to commuters.[29] The reason for more, rather than less, travel due to telecommunications is that people able to keep in regular contact with others through technologies are more likely to reinforce their physical communication, rather than replace it, although of course this happens too. Technologies enable flexible, on-the-move forms of meetingness; variable degrees of network capital; and different qualities, kinds and speeds of communication.[30]

## Homogeneity

There are many consequences for societies from further automation, particularly in the production process. In the 1960s the sociologist Herbert Marcuse made the argument that emerging automation technologies within the production and distribution complex acted not only on the kinds of commodities they made and moved, but also on the societies who came to depend upon them, worryingly homogenizing their cultures. In short, the assembly line and other mass manufacturing technologies not only shaped objects but also societies.

Marcuse joined a long line of social scientists, from the Frankfurt School (Horkheimer and Adorno) back to Karl Marx, who understood the turn to automation in production to be irreconcilable with heterogeneity. It is principally the manufacturing and logistical automation of the mid-twentieth century that for Marcuse was the indelible technology shaping culture, politics and indeed everything else in 'developed' societies, however he was at pains to insist that it was not only production and distribution that were to blame. Marcuse was by no means a technological determinist. For him, the client-side (the bit which consumers engage with) of this complex acts just as much upon the individual's needs and aspirations as the back-end (production and distribution). Marcuse was one of the first thinkers to make the link between the production process and consumer markets; he suggested that together they were creating a 'one-dimensional society' where people coerced themselves through ideologies of positive thinking into demanding homogeneity, thereby shunning choice, variety, and in the process protest and dissent.

For Marcuse, the trend of automation was not a facet of the march of progress, but rather a socio-technical phenomenon, insidiously pervading society and promoting 'unfreedom', that is, subjection to a productive [technological] apparatus.[31] Through the inculcation of social norms that depended upon the production, consumption, distribution complex a society would organize the lives of its members in a 'comfortable smooth, reasonable, democratic unfreedom', as Marcuse memorably put it.[32] What was at stake, according to Marcuse, was the sanctity of the individual's conscience no less. The curtailment of freedom was an inevitable consequence of rising standards of living through systems of mechanization and standardization that fulfilled their own prophesies of better living. The awareness of this ideology within advanced industrial civilization provided Marcuse with a blank canvas from which to launch an admirable critique of the extant consequence of what he termed the one-dimensional society: 'positive thinking' and the subjection of critical discourse through technological rationality.

There is much to credit in Marcuse's ideas, particularly in light of the present rise to domination of neoliberalism as a global orthodoxy.[33] Consumer markets and marketing are indeed part and parcel of an overarching system as the synchronization of product cycles and advertising attests: the spheres of production and circulation overlap: 'Automation indeed appears to be the great catalyst of advanced industrial society'. For Marcuse, social advancement comes at a price. Modern progress sees a turn from quality to quantity; to diminishing manual labour power in favour of machines; and the exaltation of free time for leisure, pleasure, and volunteer work.[34] As Marcuse points out, there is a catch: automation does not simply substitute leisure for manual labour. Instead, the ideology of automation percolates into people's everyday lives, automating the way they think and act.

Crucially, in Marcuse's book it is not just 'mass' production, distribution and consumption systems, but also mass information ones too that shape people's thoughts and actions. These systems are notably the global mass media, but also, increasingly, education itself, with the 'massification' of universities.[35] Of course, this homogenization thesis made sense to Marcuse as industrial automation spurred a shift to white-collar 'knowledge' professions as blue-collar manual labour receded in the mid-twentieth century. With the proliferation of information and communication technologies into the knowledge economy so too has automation had an impact upon intellectual tasks, standardizing and mechanizing these, while not necessarily making them any less onerous.

Marcuse's ideas in the *One-Dimensional Man* are undoubtedly a product of their time despite resonating in today's world no less than in the mid-twentieth century. Living in the US in the 1950s he would have been aware of a wider commentary on technological automation and its impact on so-called advanced societies. A trace of the discourse is captured in print in a 1959 edition of *Life Magazine* in a feature with the title 'Cause of Breakthrough Toward Life of Plenty'. In this case, automation brings perquisites to workers in the form of leisure time and affordable goods and profit: 'It dignifies labour by wiping out drudgery' and provides a 'windfall of inexpensive goods … to bring more comfort and enjoyment into daily life'.[36] Yet, the ultimate consequence of the shift to automation, 'its greatest boon', is the automation of knowledge work: 'thousands will be freed from boring monotonous jobs to take on more creative satisfying work, the indispensable condition for a better life'.[37] Positive thinking aside, the place of this feature alongside automobile and tobacco advertisements is a window into the kinds of industrial automation that people in the 1950s were assuming were on the near horizon: sites of work with far fewer people in far more amenable conditions, whether this was computerized clerkship in banking, a 'milling machine with a mind of its own', or unpeopled assembly lines in factories. Automation threatens both musclepower and brainpower.

Yet a further, less-positive consequence surfaces in the discourse on automation in 1964 in another *Life Magazine* feature by Ernest Havemann, portentously in the same year as the publication of *One-Dimensional Man*. Titled 'Automation and a Shrinking Work Week Bring a Real Threat to All of Us: The Emptiness of Too Much Leisure', here a more sombre appraisal of automation is given. Instead of freeing up time, people felt more harried. Moreover, encroaching upon leisure time are 'marginal chores of modern living' that appear much like work itself. More deeply, the feature reflects on the psychological consequences of leisure saturation in the lives of workers who experience guilt and anxiety from not working, so-called 'weekend neurosis'.[38]

So the standards of living that advanced industrial society affords are a trade-off as society in the process becomes 'one-dimensional'. Automation as the character of the basic productive force removes the sense of injustice and plight within manual labour and in the process renders obsolete class conflict and provides a containment of change, as industrials become administrators and workers the administrated. Next we examine how much these ideas depend on an historical notion of technological automation and analyse why the trajectory towards 'total automation' anticipated in the mid-1960s failed to materialize.

By highlighting that 3D printing is not necessarily a driver of homogeneity, since it shifts the means of production to consumers themselves, we argue that it is actually sympathetic to consumer demands for variety, choice and novelty. We draw the conclusion that far from ending up with a 'one-dimensional society', the mass manufacturing system is complicit in the emerging ubiquity of 3D printing in a 'three-dimensional' society.

Paradoxically, the latest advances in technology in mass production, distribution and consumption champion individual productivity, and more deeply, the capacity for consumers to opt-out of being consumers altogether through becoming creative, in control, producers themselves. No doubt Marcuse would be critically aware of any emergent ideals about consumer freedom apparent in commentaries on 3D printing, and in this book so are we. Similar backlashes to mass farming in the form of home gardening are also accompanied by changes in advertisement and consumer culture that do not necessarily conflict with the one-dimensional society (organic food, farm-fresh products, micro-breweries, and so on). Nevertheless, the kinds of technologies and associative social practices that are now emerging, both open source and in the marketplace, were not on Marcuse's radar in his understanding of the trajectory of progress of industrial, logistical, retail, or informational automation.

## Rapid prototyping

We now turn to a brief history of 3D printing and its emergence within the current triad. Similar to the Industrial Revolution, and the computer one in the late-twentieth century, the recent attention to 3D printing and a future systemic reconfiguration revolves around key 'heroic' innovators – these 'patent-holding, individuals-turned-entrepreneurs successfully commercialized 3D printing technology, with Scott Crump of Stratasys, Inc., and Chuck Hull of 3D Systems as prime examples' (see Figure 2.1).[39] Social network analysis of 3D printing patents

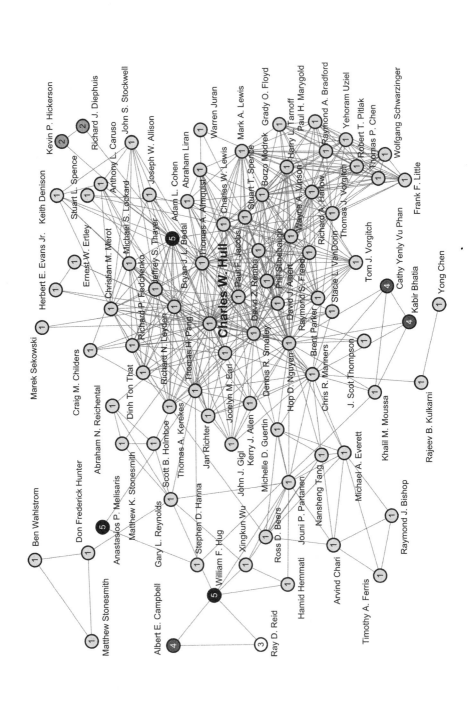

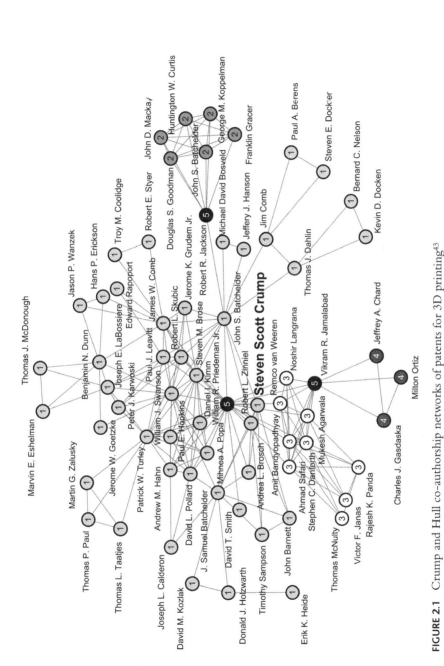

**FIGURE 2.1** Crump and Hull co-authorship networks of patents for 3D printing[43]

shows the centrality of these two individuals, in particular across organizations and institutions, as well as the many other people involved in bringing a product to market.[40]

In the late 1980s Carl Deckard and Joseph Beaman of the University of Texas (Austin) conceived of the idea of selectively sintering (heating in order to compact a solid mass without causing a 'phase transition' to a liquid) a powder deposit only a few thousandths of a millimetre thick with a laser in a chamber until a programmed shape was made.[41] A similar process, FDM, was the brainchild of Scott Crump, founder of the company Stratasys, who came upon a working prototype while attempting to manufacture a toy frog for his daughter using a computer controlled glue gun. As Crump remembers, 'I decided to create a toy froggy ... once I saw my daughter using the part ... and having gone through a number of years trying to get to market with a product, that's when the "aha" moment occurred'.[42] The use of lasers and heated nozzles to form plastic and later metal objects would prove to be a watershed for 3D printing as a method for both pre-production and production itself.

The realization that 3D printers could not only make models but could also make usable parts in plastic and metal resulted in a sea change. It was in the middle of the first decade of the 2000s that predictions began to appear in the media about 3D printing (and other similar forms of 'rapid prototyping') being a 'breakout technology' of social consequence. Increasingly commentators foresaw a potential transition from mass manufacturing to mass customization involving production runs of unique items rather than of tens of thousands of identical copies.[44] The ramifications for society of this cannot be underestimated.

One of the closest parallels in terms of social effects and systemic reconfiguration for 3D printing is two-dimensional (2D) printing. In the mid-twentieth century commercial printing took place in industrial centralized settings with many distinct technologies: woodblock floors, chases, inks, presses, types, paper rolls, binders and so on. There were also many distinct divisions of labour either 'skilled' or 'unskilled': dot-etchers, compositors, linotype operators, letterpress printers, press machinists, monotype operators, and camera operators, to name a handful of many. The introduction of digitized, decentralized desktop and office printing of various scales made these trades redundant and spaces of production just a memory.[45]

There are similarities then between 3D and 2D printing. Interestingly, 2D printing companies were forerunners of 3D printing patents, from HP to Kyocera, perhaps sensing its social significance.[46] Yet, as with most new technologies, the leading companies came from left-field. Two emerged as market leaders: 3D Systems, which recently acquired their competitor Z Corporation, and Stratasys, which recently merged with competitor Objet to create a further major corporation. Both companies were early pioneers of these innovations.

If anything is to be learned from the shift in paper printing from centralized industrial facilities to people's homes and workplaces in the late-twentieth century, there will be impacts on employment prospects, trade skills and qualifications, local and regional economies and creative endeavour. With 2D digital printing social practices also changed as people were able to rely on near to hand printers instead of handwriting for proofing, editing and correspondence. And similar to paper, 3D printer users are now able to print objects themselves on their personal machines or in office or bureau facilities – that is, specialist print shops – or even community units shared in libraries and universities.

The term '3D printing' is used to cover up to 18 different processes, although they have in common the computer-guided additive layering of materials.[47] Extrusion systems – that is, FDM – are the most popular type, with start-up companies offering models to consumers alongside industrial 'office' styled printers. In its most standard form FDM involves the heating and extrusion of thermo-plastic coils of filament wire from a 'glue-gun' through a nozzle onto a build tray. A computer guides the nozzle and the heated plastic solidifies in the form dictated by the digital design. Recent advances have expanded the ranges of materials that can be processed in FDM, from traditional polymers to organic ones and other more esoteric hybrids or 'biomaterials'.

FDM 3D printers have been used for many years in rapid prototyping of goods that would then be manufactured en masse in conventional ways. Such prototypes were also popularized by consumer home-based versions such as the RepRap, which benefited from the expiry of FDM in the 1990s, thus allowing them to become open source with their designs freely distributed. FDM 3D printers have sparked a range of products aimed at both the professional and amateur hobbyist markets. Commercial FDM 3D printers are available to buy pre-assembled (the Up! and Makerbot Replicator) that also exploit the expiry of patents.

The second, more recent, major type of 3D printing is based on binding, melting or sintering material powders. Chiefly, this is known as SLS; however, selective laser melting (SLM) and direct metal laser sintering (DMLS) are related 'breakout technologies'. SLS 3D printers allow molecular levels of detail from the fine fusion of powdered plastic, resin, nylon, steel, titanium and other materials.

There are now various machines printing many 3D shapes, the main differences being in how the layers of the print are built up as they are deposited one on top of the other as the printer releases material from a container or cartridge. There are both low-end consumer level printers available for less than a thousand pounds sterling to high-end industrial printers costing millions. As materials are stratified, so a 3D object gets to be produced. Each layer is in effect a digital slice generated through a given computer-aided design. Every next layer is added until the object is fully printed or 'manufactured', with an extruder (fused-filament), chemical agent (binder) or a laser (sintering/melting) changing the state of the material into a robust object.

A 3D printer is thus a computer-controlled device that layers materials with great accuracy and can produce objects in a range of feedstock (plastic, steel, titanium, resin, stone, ceramics, wood) similar to how a paper printer produces pages of text and images in the home or office.[48] Fine resolutions of deposits undergo a treatment (for instance, heating, sintering, melting or chemical binding) in order for the computer to 'build' additively an object in real-time.

3D printing is more technically known as 'additive' manufacturing, by contrast with most previous 'subtractive' manufacturing processes that involved cutting, drilling or bashing wood or metal or other materials. These are 'subtractive' with objects being produced identically in bulk volumes from set layouts or configurations of tools.

Instead of subtracting material, additive manufacturing uses precision techniques to build up the product layer by layer.[49] Thus 3D printing is not like other forms of industrial manufacturing. Experts are adamant that cost effectiveness is problematic to achieve for two of the major production techniques, DMLS and electron beam melting (EBM): 'High specific costs ...of material deposited respectively, are identified as a central impediment to more widespread technology adoption of such additive systems ... the observed deposition rates are not sufficient for the adoption of EBM and DMLS in high volume manufacturing applications'.[50] Hence if 3D printers are not in a position to simply slot into production facilities as they stand today, the hype around a possible systemic impact is more relevant for the social rather than the technical aspects of human-made objects. History is useful here.

Prototyping is unavoidable in the triad, as before bulk production – involving assembly line milling, moulding, cutting and so on – product designs require 'form-fit-and-function' testing to check for errors, defects, scale issues, aesthetic oversights, and unforeseen three-dimensional properties. Simply put, designs 'on the page' (or screen) do not offer industrial designers enough of a sense of what the final product will look like and how it will perform during its lifetime of use. And so rapid prototyping enters the story.

Before 3D printers came onto the prototyping scene an industrial designer, architect or engineer would need to either invest in a skilled artisan for a scale model or run the risk of producing a small batch of test subjects. Automating the modelling process with machines, so they were veritably untouched by human hands in the same fashion as bulk production methods, was a logical step in the industrial automation of the pre-production process in the late-twentieth century.[51] As summarized in the first chapter, in the late 1980s a novel technology offered manufacturers a solution to 'make it in a minute' with the latest invention of the laser and SLA.[52]

There was excitement in the late 1980s within industry circles about the curing, with light, of liquid polymers into solid objects to bypass conventional prototype model making.[53] The inventor of the process of SLA, Hull, was boosterish about the benefit of prototyping in this rapid fashion: 'By shortening the design/manufacturing cycle [it] helps companies fulfil a major goal: making a better product in a shorter amount of time'.[54] Other innovations emerged at the same time including Liquid Thermal Polymerization (LTP), similar to SLA except using an infrared laser for curing; Beam Interference Solidification (BIS), involving two laser beams to polymerize liquid resin; Solid Ground Curing (SGC), combining wax with UV light and resin; Holographic Interference Solidifaction (HIS), using a holographic image in a resin; and electrosetting (ES), using aluminium and a treated fluid.[55] 3D printing became a focus of investment and innovation to cater for the surging demand for rapid prototyping. Rapid prototyping derives from a range of 'printer' technologies allowing the local production of objects, much in the same way as 2D printers allow the generation of paper documents in offices and homes often distant from where the text has been 'designed'. Rapid prototyping can be seen as a tentative stage in a wider adoption and engagement with 3D printing throughout contemporary societies.

Many commentators during the rapid prototyping phase of development described it as a game changer in engineering. 3D printing was one of a suite of 'desktop manufacturing' technologies, alongside miniature CNC machines, that caught the attention of industry pundits.[56] The automation of prototyping was heralded as likely to become 'a standard engineering tool in the 1990s'.[57] However, as the technology progressed observers foresaw it becoming a method to transform the production process of end-user, finished products too. As 3D printing developed, so it was realized that a much wider range of shapes and materials could be produced in quantity and not just the prototypes. The 3D printing of parts or even whole objects without assembly in-house also offers some degree of protection from patent and intellectual property violations. In the early twenty-first century, global factory clusters in areas such as Shenzhen, China became infamous for their 'creative economies' producing cheaper 'non-brand' copies of products, often with the same factories tooled for the brand versions.[58] Manufacturing quickly in-house guarded against this issue somewhat.

As 3D printing developed, it came to be realized that a much wider range of shapes and materials could be produced, and not just prototypes of something else but of end-user, ready-for-market objects. An estimated 20 per cent of 3D printing is now thought to be of final products rather than the printing of prototypes.[59] 3D printers are now making their way into many industrial processes in aerospace and luxury product manufacturing where there is demand for custom features and accessories. Some examples are medical implants with novel shapes, jewellery impossible to make any other way and sportswear moulded – that is, 3D digitally scanned – to the body.[60] Engineers are 3D printing entire wings of aircraft, electric vehicle bodies, and the concrete shells of buildings. One reason for the use of 3D printing in advanced manufacturing is that the process offers designers the option of complex geometric designs that are practically impossible in other forms of manufacturing. For example, engineers at the University of Southampton made the world's first fully printed Unmanned Aerial Vehicle (UAV) in nylon.[61] And in the US Kor Ecologic and Stratasys have printed a hybrid electric car called the Urbee with a streamlined 'organic' shape.[62] A steering wheel can now be 3D printed based on a digital scan of the driver's hands. Another area where 3D printing is moving beyond rapid prototyping is in medical and health applications. The 3D printing of models of organs and prosthetics is now nearing mainstream. The 3D printing of organic transplants, such as teeth, using stem cells and organic matter, is also on the near horizon.[63]

The majority of objects 3D printed through SLA and FDM are not made as end-user products, however in the mid-2000s patent data available in the public domain showed that SLS 3D printing had shifted from simply being a form of rapid proto-typing to being an advanced method for manufacturing final objects (initially termed 'CAD-casting').[64] This shift reflected the rise of SLS 3D printing as a way for some companies in the automotive, aerospace and biotech sectors to start to apply the technique to a mass customization business model for components of aeroplanes and cars and in 'personalized' medical prostheses. We interpret the push for SLS as due to the urge in automotive to get final, custom products to market quickly and cost-effectively; the demand in aerospace for complex high-performance products in small runs; and the ready convertibility of medical imaging data for SLS 3D printing in biotech.[65]

In short, a major step in the ubiquity of 3D printing is the spread of the notion that these new technologies can be used as a manufacturing process for end-use products.[66] While FDM 3D printing is not able to produce a wide range of objects of acceptable quality for consumers used to mass manufacturing standards, SLS 3D printing does have this scope due its higher layer resolution, wider range of materi-als, and other benefits. Throughout the 2000s individual innovators become aligned with major SLS patent-owning 3D printing companies: Ingo Ederer (Voxeljet) and Rainer Hoechsmann (ExOne Company). Other recent research shows that top patent applicants, including Stratasys and Z Corp, 'have filed for patents in this area only relatively recently', obviously sensing momentum.[67] In the SLS scene, three key industries (automotive, aerospace and medical) are driving investment alongside the emergence of specific SLS 3D printing companies to facilitate bureau services to cater for growing demand. The range of displays at a global industry event such as the London 3D Print Show brings into stark relief the differences between the FDM printers of start-up and entrepreneurial companies targeting home consum-ers and hobbyist 'makers' – mostly derived from open source technologies – and professional SLA units and bureau services for rapid prototyping and advanced industrial manufacturing. The expiry of key patents in FDM and SLS are windows of opportunity for 3D printing to become ubiquitous.[68]

In 1989 a news feature on *Good Morning America* premiered a report on a radical new approach to the pre-production modelling of objects. The invention was spurred by a lack of leanness in the design and testing of objects before being sent for bulk manufacture: 'the auto industry spends twice as much money designing car parts as making them', the newscaster noted, 'both expensive and time consuming'.[69] The feature also included an interview with the president of the start-up company 3D Systems and inventor of SLA. In Hull's own words, 'I think a good way to describe it is a three-dimensional printer … in the broader sense you might say it does for engineering and manufacturing what the Xerox machine or word proces-sor or both of those do for the office environment'.[70] His colleague Raymond Freed, CEO of 3D Systems, elaborated:

> I think the technology is capable of what I call just-in-time manufacturing, which is what the world is trying to really do, which means that you would produce the part just as you need it – now we're not there yet, we've probably

got five years or more of hard research and development, but think if we could make a whole car door in less than a minute without any tooling and change it by just changing the computer model. I think we would revolutionize the way industry works.[71]

3D printing is now on the cusp of ubiquity by some accounts, however 3D printers are already reaching a 'massification' of sorts on many designers', architects' and engineers' desktops for pre-production work. Rapid prototyping became standard in industrial design, architecture and technical studios throughout the 1990s and early 2000s. The origins of 3D printing lie with the innovation of SLA. An electronic device much like a lithograph creates a stereo or three-dimensional model, usually to scale, from stock material – laser-cured, photo-reactive resin in the first instance – for testing product designs.[72] This alternative automated pre-production process was deemed 'rapid' at the time in comparison to factory retooling, transoceanic freight, or hand modelling in clay; all onerous in both cost and time.

Already at the early stage of the 1980s the inventors of 3D printing were thinking about the production of final objects and methods to make the production process more 'lean' and 'flexible' through reducing the distances involved in the production of these test models or 'rapid prototypes', which had to be produced in the approximate materials and qualities of the finished objects. One solution was SLA. An important point in the history of 3D printing is that rapid prototyping technologies overcame the shortfalls of the assembly line by allowing the pre-production process to be more lean, as redundant copies were no longer produced in mass and flexible as prototypes could be easily altered and remade before factory retooling took place.

At the low-end of SLA 3D printing are the consumer units now available from high street and online retailers with minimal set-up requirements targeted for home usage. The main uses at the low-end of the market involve the creation of custom (also termed 'bespoke') objects and parts of an experimental nature. There are a number of limitations in the 3D printing ecosystem and these particularly apply to low-end, consumer home printers: difficult-to-use design software; market-standard object finishes and qualities; expensive material feedstock; limited colours, materials and mixtures; and limits to build-tray sizes. These do not appear to be deterring commentators' enthusiastic reports of growing numbers of early adopters, despite some industry experts noting a similarity to the craze for bread machines in the 1990s – now far from a ubiquitous technology – with many similar functional elements and challenges.

While the commentary on 3D printing certainly contains a significant degree of hype there are common features in this arena that point to some or all of the elements being significant for societies in future. Despite there being an almost incomparable degree of difference between the low-end and high ends of 3D printing, a unifying trend across both is the idea that manufacturing and more generally production and consumption could converge. Central here are the early adopters and what can be termed 'maker movements' who, in experimenting with 3D printing, are driving the innovation forward. Certainly many of the uses makers put to 3D printing are hardly earth shattering – the production of novelties, that is, cake decorations or lifelike dolls, being a case in point. Yet even a cursory survey of online and free-to-use design repositories reveals some objects of more important social use.

At the low-end of the scale of 3D printers are very cheaply built open source machines such as the Reprap. These print objects with plastic filament wire and use open source motherboards. There are similar machines already available as a kit or pre-built by start-up companies such as Makerbot. It is surprising how much low-end printers can do. Objects with moving parts (cogs, gears and wheels) can be printed pre-assembled. Scans of faces and other detailed images are rendered in excellent detail depending upon the printed layer size – the finer the better. Innovation at this low-end of the market for 3D printers is directed towards even finer print-layers; printers with mixed materials and colours; intuitive user interfaces and design-by-wire software packages; the embedding of circuits into designs; so-called 'self-replication' through the printing of printer parts; universal CAD file formats that can be used across software packages and printer types; and the creation of online repositories with inventories of designs printable at a low cost and with different materials by non-technical users.

## Conclusion

It can be argued here that technologies do not simply substitute for other ones, instead 'new' technologies enmesh with 'old' ones, bringing many uncertainties into transitions.[73] An example is that early steamships at the turn of the twentieth century sported both sails and steam chimneys.[74] Technologies do not operate in a vacuum. Instead they evolve multi-directionally – sometimes unsuccessfully or with a delay – as historical artefacts in relation to social groups. Individuals and societies construct meanings and uses from and with them, and in turn shape systems through policies, planning, and the trends in norms that are set by social practices.[75] In short, technologies are socially produced in a variety of social circumstances.[76] In some cases technologies, such as the automobile, are an element within social change at the very largest scale.

Going back to Adam Smith's pin factory we can inquire: what is a pin for? In the twenty-first century household pins – invariably 'safety' ones – hold diapers together or modify ill-fitting clothes: they allow users to repair or customize objects themselves. However, they do this in an inefficient manner and the current system of manufacturing objects offers an alternative. Instead of being reused, diapers are bought in bulk and disposed of after single use. Clothes too are worn and once requiring repair or no longer fitting their owner are more often than not disposed of in the same fashion as disposable nappies. Most people today would be hard put to remember the last time they utilized a safety pin beyond perhaps as a punk accoutrement.

With 3D printing an entirely different vision is offered to either the inefficient repair or customization of pins and the rapid replacement of mass manufacturing of objects. With 3D printers users are able to make clothes that fit perfectly to their bodies. Similarly, users are able to repair objects themselves with identical parts or print the object again anew regardless of whether it is still on the market or not.

The question of whether 3D printing is or is not an appropriate technology for producing pins can be put another way: what is it good for? 3D printing is unlikely to be economical for mass-consumed objects in the foreseeable future unless it disrupts, or reconfigures, the current system in terms of different degrees of value and competitive advantage. As we have seen in this chapter already 3D printing is an established method for adding value in pre-production model making: rapid prototyping. However, there are other contributions it can make as analysts are now realizing.[77] In later chapters the two uncertainties of engagement and openness shape the thinking in this book on the social consequences of 3D printing by defining four futures foreseeing mass adoption.

In the next chapter we understand the 3D system in reference to the excitement it attracts in the public domain and media. There are parallels with how production was done prior to the Industrial Revolution, which reached its apex late in the nineteenth century. With the shift to industrial manufacturing, many crafts became less significant as unskilled workers entered factories. Here, industrial fabrication was performed by machines involving highly routinized operations and assembly by industrial workers based upon a complex division of labour, famously elaborated in Adam Smith's eighteenth-century account of a pin factory, resulting in economies of scale. A 3D printer could manufacture a pin with moving parts and no human assembly whatsoever, however that is not all it will do in a reconfigured system of production, distribution and consumption in an entirely different society from today's.

## Notes

1 No Author, 'The City of the Future', 1935 British Pathé. Accessed 1 October 2015. https://youtu.be/UZUMo_QYbB0

2 Popular Science Monthly, 'Wonder City You May Live to See May Solve Congestion Problems'. *Popular Science Monthly*. New York, August 1925. p. 41. https://books.google.com.au/books?id=YScDAAAAMBAJ&lpg=PA40&dq=%E2%80%98Wonder%20City%20You%20May%20Live%20to%20See%E2%80%99%201925&pg=PP1#v=onepage&q&f=false

3 N. Hopkinson, R.J.M. Hague and P.M. Dickens, 'Introduction to Rapid Manufacturing'. In *Rapid Manufacturing: An Industrial Revolution for the Digital Age*, edited by Hopkinson and Hague, 1–4. Chichester: John Wiley & Sons, 2006.

4 D.E. Nye, *America's Assembly Line*. Cambridge, MA: MIT Press, 2013.

5 A. Smith, *Wealth of Nations*. London: ElecBook, 2001(1766). p. 18.

6 F.W. Geels, 'Major System Change through Stepwise Reconfiguration: A Multi-Level Analysis of the Transformation of American Factory Production (1850–1930)'. *Technology in Society* 28, no. 4 (2006): 445–76, doi:10.1016/j.techsoc.2006.09.006

7 P. Macnaghten and R. Owen, 'Environmental Science: Good Governance for Geoengineering'. *Nature* 479, no. 7373 (2011): 293–93.

8 S.D.N. Graham, 'Software-Sorted Geographies'. *Progress in Human Geography* 29, no. 5 (2005): 562–80, doi:10.1191/0309132505ph568oa

9 E. Wyly, 'Automated (Post)Positivism'. *Urban Geography* 35, no. 5 (2014): 669–90, doi:10.1080/02723638.2014.923143

10 N. Castree, A. Rogers and R. Kitchin, *A Dictionary of Human Geography*. Oxford: OUP, 2013. p. 505.

11 C. Murray, 'Women's World Cup: Is Artificial Turf to Blame for a Lack of Goals?'. 2015 Guardian News and Media Limited. Accessed 29 July 2015. http://www.theguardian.com/football/2015/jun/18/womens-world-cup-artificial-turf

12 N. Zuberi, 'Is This the Future? Black Music and Technology Discourse'. *Science-Fiction Studies* 34, no. 2 (2007): 283–300.

13 D. MacKenzie and J. Pablo Pardo-Guerra, 'Insurgent Capitalism: Island, Bricolage and the Re-Making of Finance'. *Economy and Society* 43, no. 2 (2014): 153–82, doi:10.1080/03085147.2014.881597.

14 D.E. Nye, *Technology Matters: Questions to Live With*. Cambridge, MA: MIT Press, 2006.

15 M. Weber, *The Protestant Ethic and the Spirit of Capitalism*. London: George Allen and Unwin, 1976.

16 The Economist, 'Making the Future'. *The Economist*, 21 April 2012. Accessed 7 July 2012, http://www.economist.com/node/21552897

17 J. Meynaud, *Technocracy*. London: Faber and Faber, 1968.

18 The Economist, 'The Onrushing Wave'. *The Economist*, 18 January 2014. Accessed 15 June 2015, http://www.economist.com/news/briefing/21594264-previous-technological-innovation-has-always-delivered-more-long-run-employment-not-less

19 D.E. Nye, *Technology Matters : Questions to Live With*. Cambridge, MA: MIT Press, 2006. p. 165.

20 S. Lash and J. Urry, *The End of Organized Capitalism*. Cambridge: Polity Press, 1987.

21 A.S. Blinder, 'Offshoring: The Next Industrial Revolution?'. *Foreign Affairs*, 1 March 2006. Accessed 16 May 2012, http://www.foreignaffairs.com/articles/61514/alan-s-blinder/offshoring-the-next-industrial-revolution

22 I. Oshri, J. Kotlarsky and L.P. Willcocks, *The Handbook of Global Outsourcing and Offshoring*. Basingstoke: Palgrave Macmillan, 2009.

23 D. Cowen, *The Deadly Life of Logistics: Mapping Violence in Global Trade*. Minnesota: University of Minnesota Press, 2014.

24 P.L. Mokhtarian, 'An Empirical Evaluation of the Travel Impacts of Teleconferencing'. *Transportation Research Part A: General* 22, no. 4 (1988): 283–89, doi:10.1016/0191-2607(88)90006-4; P.L. Mokhtarian and I. Salomon, 'Emerging Travel Patterns: Do Telecommunications Make a Difference?'. In *Perpetual Motion: Travel Behaviour Research Opportunities and Application Challenges*, edited by Mahmassani, 143–82. Oxford: Elsevier Science Ltd., 2002.

25 R. Sims, R. Schaeffer, F. Creutzig, X. Cruz-Núñez, M. D'Agosto, D. Dimitriu, M. J. Figueroa Meza, L. Fulton, S. Kobayashi, O. Lah, A. McKinnon, P. Newman, M. Ouyang, J.J. Schauer, D. Sperling and G. Tiwari, 'Transport'. In *Climate Change 2014: Mitigation of Climate Change. Contribution of Working Group III to the Fifth Assessment Report of the Intergovernmental Panel on Climate Change*, edited by Edenhofer, Pichs-Madruga, Sokona, Farahani, Kadner, Seyboth, Adler, Baum, Brunner, Eickemeier, Kriemann, Savolainen, Schlömer, von Stechow, Zwickel and Minx, 599–670. Cambridge: Cambridge University Press, 2014.

26 J. Kietzmann, L. Pitt and P. Berthon, 'Disruptions, Decisions, and Destinations: Enter the Age of 3-D Printing and Additive Manufacturing'. *Business Horizons* 58, no. 2 (2015): 209–15, doi:10.1016/j.bushor.2014.11.005

27 D.R. Gress and R.V. Kalafsky, 'Geographies of Production in 3D: Theoretical and Research Implications Stemming from Additive Manufacturing'. *Geoforum* 60 (2015): 43–52, doi:10.1016/j.geoforum.2015.01.003

28 P.L. Mokhtarian, 'An Empirical Evaluation of the Travel Impacts of Teleconferencing'. *Transportation Research Part A: General* 22, no. 4 (1988): 283–89, doi:10.1016/0191-2607(88)90006-4

29 F. Dal Fiore, P.L. Mokhtarian, I. Salomon and M.E. Singer, '"Nomads at Last"? A Set of Perspectives on How Mobile Technology May Affect Travel'. *Journal of Transport Geography* 41 (2014): 97–106, doi:10.1016/j.jtrangeo.2014.08.014

30 J. Urry, 'Social Networks, Travel and Talk'. *The British Journal of Sociology*, 54, no. 2 (2003): 155–75, doi:10.1080/0007131032000080186

31 H. Marcuse, *One Dimensional Man*, 3rd edn. Great Britain: First Sphere Books, 1970. p. 41.

32 Ibid., p.13.

33 J. Urry, *Climate Change and Society*. Cambridge: Polity Press, 2011.

34 H. Marcuse, *One Dimensional Man*, 3rd edn. Great Britain: First Sphere Books, 1970. p. 44.

35 R. Barcan, *Academic Life and Labour in the New University: Hope and Other Choices*. Farnham: Ashgate Publishing Ltd, 2013.

36 Life Magazine, 'Cause of Breakthrough toward Life of Plenty'. *Life Magazine*, Special Issue: The Good Life, 28 December 1959, p. 36.

37 Ibid., p. 36.

38 E. Havemann, 'Automation and a Shrinking Work Week Bring a Real Threat to All of Us: The Emptiness of Too Much Leisure'. *Life Magazine*, 14 February 1964.

39 C.M. McNulty, N. Arnas and T.A. Campbell, 'Toward the Printed World: Additive Manufacturing and Implications for National Security'. *Defense Horizons*, 73 (2012): 1–16. p. 11.

40 T. Birtchnell, G. Viry and J. Urry, 'Elite Formation in the Third Industrial Revolution'. In *Elite Mobilities*, edited by Birtchnell and Caletrío, 62–77. Abingdon: Routledge, 2013.

41 F. Lavoie, 'Is 'Desktop Manufacturing' for You?'. *American Machinist & Automated Manufacturing* 133, no. 3 (1989): 61–3.

42 No Author, 'Stratasys Takes 3D Printing Mainstream', 2011. Accessed 23 January 2011, http://www.kare11.com/news/career/article/901211/376/Stratasys-takes-3D-printing-mainstream

43 T. Birtchnell, G. Viry and J. Urry, 'Elite Formation in the Third Industrial Revolution'. In *Elite Mobilities*, edited by Birtchnell and Caletrío, 62–77. Abingdon: Routledge, 2013. pp. 70–1.

44 Business Wire, 'Rapid Prototyping Makes Technologies Affordable by Enabling Mass Customization', *Business Wire*. Palo Alto: Frost & Sullivan, 13 July 2004. No pagination. Accessed 3 January 2015.

45 J.A. Stein, 'The Co-Construction of Spatial Memory'. *Fabrications* 24, no. 2 (2014): 178–97, doi:10.1080/10331867.2014.961222

46 S. Shankland, 'HP Joining 3D Printer Market with Stratasys Deal'. *CNET News*, 2010. Accessed 20 August 2012, http://news.cnet.com/8301-30685_3-10436841-264.html

47 N. Hopkinson, R.J.M. Hague and P.M. Dickens, eds, *Rapid Manufacturing: An Industrial Revolution for the Digital Age*. Chichester: Wiley, 2006.

48 D. Sieberg, BBC News – World News America: *3D Printing Creates 'Something out of Nothing*, 2010.

49 Good Morning America, 'Science Segment with 3D Systems'. YouTube, 1989. Accessed 5 September 2012, http://www.youtube.com/watch?v=NpRDuJ5YgoQ

50 M. Baumers, P. Dickens, C. Tuck and R. Hague, 'The Cost of Additive Manufacturing: Machine Productivity, Economies of Scale and Technology-Push'. *Technological Forecasting and Social Change*, 102 (2016): 193–201, doi:10.1016/j.techfore.2015.02.015. 193

51 L. Wood, 'Rapid Prototyping: Uphill, but Moving'. *Manufacturing Systems* 8, no. 12 (1990): 14–18.

52 S.J. Muraski, 'Make It in a Minute'. *Machine Design* 62, no. 3 (1990): 127–32.

53 B. Kellock, 'Excitement of Technology Trends'. *Machinery and Production Engineering* 147, no. 3767 (1989).

54 C. Hull, 'Stereolithography: Plastic Prototypes from Cad Data without Tooling'. *Modern Casting* 78, no. 8 (1988). p. 38.

55 D.T. Pham and R.S. Gault, 'A Comparison of Rapid Prototyping Technologies'. *International Journal of Machine Tools and Manufacture*, 38, no. 10–11 (1998): 1257–87, doi:10.1016/S0890-6955(97)00137-5

56 F. Lavoie, 'Is "Desktop Manufacturing" for You?'. *American Machinist & Automated Manufacturing* 133, no. 3 (1989): 61–3.

57  D. Deitz, 'Stereolithography Automates Prototyping'. *Mechanical Engineering*, 112, no. 2 (1990): 35–9.

58  M. Keane and E. Zhao, 'Renegades on the Frontier of Innovation: The Shanzhai Grassroots Communities of Shenzhen in China's Creative Economy'. *Eurasian Geography and Economics* 53, no. 2 (2012): 216–30, doi:10.2747/1539-7216.53.2.216

59  R.B. Kross, 'How 3D Printing Will Change Absolutely Everything It Touches', 2011 Forbes. Accessed 5 August 2011. http://www.forbes.com/sites/ciocentral/2011/08/17/how-3d-printing-will-change-absolutely-everything-it-touches/

60  R. Nayak, R. Padhye, L. Wang, K. Chatterjee and S. Gupta, 'The Role of Mass Customisation in the Apparel Industry'. *International Journal of Fashion Design, Technology and Education* 8, no. 2 (2015): 162–72, doi:10.1080/17543266.2015.1045041

61  P. Marks, '3D Printing: The World's First Printed Plane', 2011. Accessed 23 August 2012, http://www.newscientist.com/article/dn20737-3d-printing-the-worlds-first-printed-plane.html

62  S.S. Nathan, 'Solid Progress: 3D Printing Technology is Being Applied in Some Unexpected and Unusual Areas'. *Engineer*, no. MAR (2013): 26–9.

63  D. Lupton, 'Fabricated Data Bodies: Reflections on 3D Printed Digital Body Objects in Medical and Health Domains'. *Social Theory & Health* 13, no. 2 (2015): 99–115, doi:10.1057/sth.2015.3

64  M. Freeman, 'Build Parts by Printing'. *Design Engineering*. London: Centaur Communications Limited, 1997.

65  J. Bhattacharjya, S. Tripathi, A. Taylor, M. Taylor and D. Walters, 'Additive Manufacturing: Current Status and Future Prospects'. In *Collaborative Systems for Smart Networked Environments*, edited by Camarinha-Matos and Afsarmanesh, 365–72. Berlin/Heidelberg: Springer, 2014. p. 366.

66  N. Hopkinson and P. Dickens, 'Analysis of Rapid Manufacturing – Using Layer Manufacturing Processes for Production'. *Proceedings of the Institution of Mechanical Engineers, Part C: Journal of Mechanical Engineering Science* 217, no. 1 (2003): 31–9, doi:10.1243/095440603762554596

67  Intellectual Property Office, *3D Printing: A Patent Overview*. Newport: UK Government, 2013.

68  T. Birtchnell, G. Viry and J. Urry, 'Elite Formation in the Third Industrial Revolution'. In *Elite Mobilities*, edited by Birtchnell and Caletrío, 62–77. Abingdon: Routledge, 2013.

69  Good Morning America, 'Science Segment with 3D Systems', YouTube, 1989. Accessed 5 September 2012. http://www.youtube.com/watch?v=NpRDuJ5YgoQ

70  Ibid.

71  Ibid.

72  S. Bradshaw, A. Bowyer and P. Haufe, 'The Intellectual Property Implications of Low-Cost 3D Printing'. *ScriptEd* 7, no. 1 (2010), doi:10.2966/scrip. 070110.5. 29

73  D.E. Nye, *Technology Matters: Questions to Live With*. Cambridge, MA: MIT Press, 2006.

74  F.W. Geels, 'Technological Transitions as Evolutionary Reconfiguration Processes: A Multi-Level Perspective and a Case-Study'. *Research Policy* 31, no. 8–9 (2002): 1257–74.

75  E. Shove, M. Pantzar and M. Watson, *The Dynamics of Social Practice: Everyday Life and How It Changes*. London: Sage Publications Limited, 2012.

76  T.J. Pinch and W.E. Bijker, 'The Social Construction of Facts and Artefacts: Or How the Sociology of Science and the Sociology of Technology Might Benefit Each Other'. *Social Studies of Science* 14, no. 3 (1984): 399–441, doi:10.2307/285355

77  T. Rayna and L. Striukova, 'From Rapid Prototyping to Home Fabrication: How 3D Printing is Changing Business Model Innovation'. *Technological Forecasting and Social Change* 102 (2016): 214–24, doi:10.1016/j.techfore.2015.07.023

# 3

# THE 3D SYSTEM

## Introduction

In the previous chapter we demonstrated how 3D printing is neither simply 'about' rapid prototyping nor 'just' a technical development for engineers and scientists to concern themselves with; there are social ramifications and consequences only now being realized by citizen-consumers, policymakers and social commentators. Guiding this attention are other developments on the interface between technology and society. Many uses of technology reach a level of adoption understood to be 'ubiquitous' – appearing everywhere and an aspect of social transformation. The car is now ubiquitous globally. So is the mobile phone. The Internet is also understood as worldwide, however its changes are not tangible in the biosphere necessarily as they are informational and communicative.

The system innovation 3D printing will be a part of will span both the physical and the virtual worlds. Cyberspace offers an instance where social reality stretches beyond the physical world, where people's lives revolve around the tangible and intangible.[1] Notwithstanding the relatively small number of users respective to the world's population there are many incongruities between life lived in the flesh and on the screen, which makes the presence of a parallel virtual world unsettling in terms of the social effects, for instance in 'virtual adultery'.[2] Mobile Internet access is also understood now to be 'ubiquitous', at least in the EU.[3] In Australia many street phones are being replaced with wireless hotspots using the same infrastructure as telephone 'boxes', baffling for those without a smart phone or portable device. What these examples show is that once an innovation reaches a certain level of ubiquity societies begin to change as people adapt technologies to their wants, uses and practices. Hence the ubiquity of 3D printing will most likely have a similar effect of inculcating social change across physical and virtual worlds.

Such a suggestion is now popular in commentary on the innovation of 3D printing: 'Personal manufacturing technologies will profoundly impact how we design, make, transport, and consume physical products'.[4] One major change is the affordance of uniqueness. At root 3D printing offers the personal control over the production of objects (classically known as 'the means of production'), a phenomenon that declined dramatically throughout the world in the twentieth century and particularly in the Global North. A growing sense that 3D printing is reaching ubiquity is also due to its success in a range of applications from medicine to art. Those who use the devices for technical operations realize the degree to which '3D printing is ideal for making one of a kind items at cost-effective prices' in wider manufacturing.[5] In this chapter we argue that the hype around 3D printing as a significant innovation is missing a key point. 3D printing is not only a technological innovation, but also a social one. We maintain it is necessary to turn to a sociological account of 3D printing in order to understand fully its significance for the future.

That debate around 3D printing's future ubiquity needs be held in the domain of social science and not just engineering or technological commentary might be surprising to many observers. However, as those people who are aware of the limitations of 3D printing know, it is unlikely to disrupt the capitalist triad of global systems of assembly line manufacturing, containerized freight and brand-informed marketing as it stands, unless society changes in some important ways. In the later chapters of this book a number of scenarios are forecast with the following trends in mind. First, consumers must face a transition in their social practices to become producers. Second, corporate capitalism must face a transition in order to cater to the ubiquity of variety and choice in object access. Finally, governments must transition to provide services, infrastructure and support to facilitate societies that have, in the Global North at least, begun to make objects again themselves.

So if 3D printers become ubiquitous in a similar fashion to the personal computer (and the laptop, smart phone and tablet) then this will substitute for, augment or increase degrees of mobility in society. What can be expected in a world of low corporatization are freely available printers, open source designs, and possibly open source and printable (so-called self-replicating) printers. In a world with a high engagement of individuals in the printing process there are likely to be peer production networks, a lack of safety and standardization unless regulated by government, and corporations seeking markets elsewhere – most likely in digital rights management, insurance services, energy supplies and other utilities, recycling and disposal, and in materials and resources provision and procurement.

As we have so far established, it is not enough to understand the current methods by which people procure objects through industrial manufacturing technologies such as the assembly line in isolation. Without mass distribution and mass consumption, there would be little profit to be made in mass production. The ubiquity of 3D printing throughout society is a non-negotiable aspect of a sociotechnical transition. CEO Cathy Lewis of company The Desktop Factory asserts that their goal is one day to make 3D printing as common in offices, factories, schools and homes as laser printers are today. There are a number of trends converging in 3D printing, which seem to be changing the rules of the game: the cost of the printers is dropping dramatically, indicating economies of scale and rapid innovation; printer design files are beginning to be stored, shared and sold; and the material base is expanding to include ceramics, metal alloys and even food.[6] Moreover, 3D model creation is being 'democratized' through alternatives to traditional CAD programs that use visualizations and templates. The ubiquity of 3D printing is also exciting venture capitalists and fuelling investment. And there is even scope to print fully assembled gadgets with multiple materials, different colours, embedded electronics and moving parts in the near horizon.[7] So 3D printers could be 'network technologies' connected to online repositories of designs downloadable in any location.[8] This could be in the home, the high street, the community centre, or the office. Each of these spaces, or combinations of them, will have distinct implications for society and transportation and whichever dominates will set the tempo of its world. Each space will involve different implications for transport patterns.

As the sociologist Gerald Davis summarizes, it is already possible to imagine 'equipping every town with a high-end do-it-yourself (DIY) facility capable of producing products from scratch based on digital designs' – from furniture and prosthetic limbs to replacement auto parts there are already many possibilities.[9] And this would entail a dramatic 're-imagining' of the corporation with large-scale production again taking place close to company headquarters if not by consumers themselves, that is if corporations do not disaggregate beyond any semblance of today's organizations.[10] Factories with 3D printing technologies will be distributable near consumers because the cost of setting up a 3D printer is the same whether it makes one item or many different items. The possibility for 'real-time mass customization' consequently blends the shopping and making experience.[11] Overall there are many possibilities for a much greater localization of manufacturing – for some non-critical products the capacity to scan the object and then make endless copies (an 'infinite aisle') by or near consumers would produce large cost savings and reduce transport-related emissions and oil use, assuming that roughly the same number of products is being manufactured worldwide.[12]

In this chapter we examine the kind of system that could emerge with 3D printing's ubiquity. We then go on to consider the difference between other machine automation trends and this one. We then critically debate the social implications of the decentralization and distribution of the means of production for work and employment and society more widely.

In deliberation on this system it is vital to note that many users of 3D printers will not be passive consumers, but 'makers' who design and craft their own objects relative to their identities. In consequence manufacturing close to the consumer via 'direct digital manufacturing' could eliminate or augment many stages in the existing system: excess raw materials, logistics, freight and inventories.[13] And it is not only retail stores that may change, spaces such as libraries and bookstores would also experience a transition.

Take the example of the library. Data handlers and knowledge providers are transforming into places for networking, meeting and collaborating with peers as well as for accessing information with the trend for people to engage more readily with data. The library is not just a repository for collections of manuscripts, but a community space where information can be retrieved and shared. While the book is the traditional vessel for information this is no longer the case. With the mass digitization of text the function of the library is changing, along with the space itself. As shelves are being removed in favour of workspaces and repositories are being automated and barcoded for rapid retrieval of information, the library is becoming a place to go for the materialization of digital data, shared through site-specific books, scans, photocopies, protected files, e-readers, computers, micro-film units and cinema displays.[14] It is a space to use data, and equipment that enables data, unavailable from elsewhere. In light of this changing functionality some innovative centres and libraries are experimenting with 3D printing in the same way libraries and community centres were early adopters of photocopiers and paper printers.[15] On top of being the place to go for Internet access, copying, scanning and printing and accessing digital collections, 'the library would become a "creative space for making things"'.[16] What this does is offer a space for people to experiment in their own free time without necessarily being concerned about time or budget. 3D printers and the resources they use will be integrated into the organizational logistics of the space in the same way a library manages the delivery, storage and procurement of its collection.[17] Libraries and similar facilities might aggregate both the factories and shops of the future.

In the next section we consider how the current triad of production, distribution and consumption follows on from a spatial shift of the means of production from the Global North to South in the 'offshoring' of production, transoceanic distribution and temporal shift in lean approaches to management, manufacturing and logistics.[18] In the second half of the twentieth century manufacturing was offshored (or outsourced) from the US and other major economies, and this was to all intents and purposes a socio-technical transition that involved the 'migration of jobs, but not the people who perform them, from rich countries to poor ones'.[19] In support of offshoring, governments use policymaking to promote international trade and reduce political, legal and geographical hurdles to the movement and sale of commodities.[20] Free trade agreements between governments have reinforced the push for 'comparative advantages' in the current regime because of the 'cost advantages of producing certain goods and services for overseas trade'.[21] Free trade policies include the North American Free Trade Agreement (NAFTA) and the political lobbying of the World Trade Organization (WTO) to limit trade barriers. Unfortunately, comparative advantages involve low-paid, non-unionized, invariably female, labour cohorts; vast emissions of GHGs and uses of energy for transportation; and much waste and cultural homogenization in consumer marketing and branding.[22]

## The triad

First there is the production system. As the 'workshop of the world', China represents the core of the current regime of 'supply chain capitalism' and the region has accumulated considerable economic and political power as a consequence.[23] A 'peculiar' pattern is evident in the current regime that involves global production networks joining low-cost locations to high ones.[24] Industry invokes production and distribution circuits and networks of interaction across uneven geographies and huge inequalities to make profit.[25] The process has been termed 'deindustrialization' or 'post-Fordism' in some quarters.[26] A declining manufacturing sector and a growing service sector in the Global North and consequent employment uncertainty and urban transformation are its hallmarks, due to sustained efforts to continue economic growth indefinitely through ever faster supply chains across ever greater distances with a consequent deterioration of unionization.[27]

Second there is the distribution system. The standardization of the shipping container technology and haulage infrastructure across freight modes (rail, road and sea) in the mid-twentieth century was a precursor to globalization as well as the deindustrialization of America and other regions in the Global North (UK, Australia).[28] The multi-modal container is part of the logic of integration that depends on uniformity of infrastructure and design conformity within tightly managed, securitized spaces to overcome regional development disparities in Tianjin, China for instance.[29] Driven by advances in information and telecommunications technologies transoceanic shipping and logistics gradually turned into an applied science.[30] The industry of logistics, which simply put is the management of the movement of things, was thought to have a value of US\$3.9 trillion by 2013.[31] Freight now involves the continuous monitoring of point-of-sales data, inventory, worker routines, and weather and traffic patterns.[32] Massive bulk retailers like Walmart depend on the container revolution for the cost of shipping goods in containers being between 1 and 2 per cent of retail value, 90 per cent less than before containerization.[33]

Third there is the consumption system. The spread of consumerism throughout the world has been documented as the steady provision of micro-geographies of services in urban space.[34] These services cater to modern identities condensing in the act of repeatedly purchasing commodities as a form of culture in their own right and different from 'high' forms: religion, art, music and so on.[35] The idea of planned obsolescence is that product lifespans can be reduced through design and material innovation so that user preferences are for 'more, better, faster'.[36] The innovation produces products with a 'shorter than economically desirable useful life' despite rational customers having full information about the product.[37] More deeply, the logic of planned obsolescence lends economic continuity to markets regardless of the cycles of waste that such on-going production creates in quantities of equal scale.[38] Yet interestingly, planned obsolescence is not a universally inevitable phenomenon, having an unequal affect on user preferences around the world, particularly on the edge of global markets, with products in the USA ending their lives earlier than those in Brazil and India.[39]

In the next section we examine the underlying logic of flexibility in the existing system, which ushered in 3D printing as a solution to friction, inefficiency and uncertainty due to time and space constraints. The third section's inquiry looks at how time-space compression involves a reconfiguration of the triad and a window of opportunity for 3D printing as a ubiquitous feature of modern societies. The three final sections then assess the possible impacts on societies in terms of production, distribution and consumption. These include the reshoring of the means of production, the decoupling of transport from object procurement and alternative kinds of markets – blended retail. 3D printing is set to upstage all three aspects of the triad and there is much scope for a front of stage role in a socio-technical transition.

## Time-space compression

Throughout the recent history of capital accumulation there has been a concerted effort towards 'time-space compression'.[40] In the twenty-first century the automation of pre-production through 3D printing is being applied to similar issues in production: time, distance and other uncertainties introduced from global production networks. The pressures that foreshadow 3D printing's future ubiquity are due to multiple instances of friction driving innovation towards 'rapid production' technologies that offer 'unlimited design flexibility without the tooling costs'.[41] Not only time but space too 'is money'.[42] All profit-makers within capitalist markets are compelled to reduce the costs incurred in the labour and transit times associated with the conversion of a raw resource into a finished product, however since the Industrial Revolution the space resources and products travel over has gone in the other direction: it has increased. As counter-intuitive as it may appear, throughout the twentieth century the spaces of production grew rather than shrank in order to facilitate the compression of time. Rather than produce objects close to where they would be consumed, it became more profitable to produce them faster at greater distances in places where comparative advantages prevailed due to economies of scale and global inequalities of various kinds. In short, time compression trumped space compression.

Few in the world would argue that current global inequalities in incomes, standards of living, energy use, environmental regulations, labour laws, modern conveniences and other features of the 'developed world' should remain in perpetuity. Even exponents of neoliberalism would acknowledge that global progress will be achieved through a 'rising tide that lifts all boats': the richer those currently benefitting from the current system get, the closer aspirants of inclusion into the system will be. With rapid prototyping, 3D printers surfaced in part as a correction of the discontinuity between time and space compression. Avoiding the costs associated with both time and space became a paramount concern.

What the progression from rapid prototyping to advanced manufacturing demonstrates is that 3D printing is progressing towards ubiquity within the current production-distribution-consumption system due to the social and economic drive for 'time-space compression', manifesting from the 'jungle law' of neoliberal capitalism, rather than technical evolution per se.[43] The desire for profit from 'leanness' is a result of investment and innovation throughout the twentieth century in 'just-in-time' (JIT) management, logistics and manufacturing.[44]

It arises that as a method of 'personal manufacturing' 3D printing has the scope to impact upon this pressure for time-space compression even more.[45] As we surmised in the first chapter of this book, the process disrupts how energy, resources, the environment and labour are utilized globally in mass manufacturing.[46] The demand for, and hype around, 3D printing are in part due to landscape pressure on the globalized and multifaceted system of international trade across a range of dimensions, summarized next.

Time-space compression was a trend that had its origin in managerial ideology and corporate culture, inspired by Japanese companies such as Toyota, who used Japan's lack of material resources to their advantage to become 'lean' in their processes.[47] Original equipment manufacturers reduced waste and enforced, in their parts suppliers, standardized components to eliminate 'dead stock' in unused inventories and create faster production times from efficiencies in freight. An unforeseen consequence of the demand for minimal disruption and maximum speed from parts subcontractors was that factory plant owners preferred regional relationships rather than extra-regional ones and so investment (in the automotive industry for instance) went east to Asia as did labour.[48]

Both space and time are relevant to 3D printing because the process's dominant qualities are a reduction in production times and a reduction in production distances. Since the 1980s JIT's influence on the current socio-technical regime has been to reduce waste, minimize inventories, and encourage workers to be more efficient.[49] JIT then had consequences across the dimensions of the current socio-technical regime. We summarize these different aspects before turning to the emergence of rapid prototyping within this regime partly due to this turn to time-space compression.

While manageable for single-firm activity with close geographical proximity to their parts suppliers, time-space compression presented many challenges for global sourcing dominated by global production networks of suppliers and subcontractors. Strategies to implement JIT globally included 'shipment consolidation with other overseas suppliers, establishment of distribution centres by the seller in buyer territory, buyer warehousing on consignment basis and the use of specialized consolidators'.[50] The term 'integration' arose in order to describe the implementation of JIT to the flow from raw materials through manufacturing, distribution centres and finally to retailers via digital information and communication technologies and freight transport.[51] The rise of online systems (Ebay, Amazon) allowed the same kind of 'leanness' to occur in post-manufacturing that the JIT ideology achieved in pre-manufacturing supply chains.[52] Yet an unintended consequence of this shift was in the growth of pressure from consumers for time-space compression.[53]

Certainly revolutionary, time-space compression in local and global sourcing, and in delivery, was not judged to be a true socio-technical transition, rather it was an internal development that placed pressure on the existing regime: 'lean production can hardly be considered as an alternative to mass production, as its proponents suggest, but is on the contrary extending the life of mass production methods'.[54] With 3D printing's capacity to actually remove the 'friction' between the finished product and the consumer, while at the time offering alternatives to the demand for making bulk numbers of identical items, an 'industrial revolution' has become a subject of conjecture, enabling 'companies to operate with little or no unsold finished goods inventory'.[55] 3D printing arose in response to these pressures for time-space compression, creating conditions for a new configuration of novelties to play a much greater role in society than in rapid prototyping.

The pressures for leanness are fomenting considerable attention and investment in 3D printing because it promises more efficient use of resources, energy and waste; so-called zero-length supply chains; more aware, democratic and sustainable kinds of consumerism due to the possibilities afforded to those who want to craft, customize or invent their own objects; and even perhaps the return, or reshoring, of manufacturing to the Global North, thus reversing the trend of offshoring, which has accelerated in many post-industrial countries in the twenty-first century.[56] Of course, these are all speculative interpretations of what a socio-technical transition in 3D printing might be like and very different futures could instead emerge with unpredictable consequences for social inequality, the environment, consumer 'freedom', and regional economic growth.

The ultimate outcome of this pressure for time-space compression is now being understood as the 'end of mass production', due to the rise of 'tailored manufacturing', allowing greater customization and engagement by consumers.[57] We argue that 3D printing has co-evolved with the functions it has come to serve in the current regime through its ubiquity in rapid prototyping in response to the demand for time-space compression. In the next section, we move to the emergence of rapid prototyping beyond the niche-level into the mainstream regime.

## Reshoring

In this section we examine the reshoring of manufacturing. Manufacturing since the 1970s has thus been 'offshored' from major production centres to Mexico, China, India, Vietnam and other countries offering low incomes for assembly lines, using local energy, mostly coal, and bearing the environmental costs. The offshoring of production and manufacturing was a key factor in the growing power of the Brazil, Russia, India and China (BRIC) economies and the rapidly escalating emissions of greenhouse gases (GHGs) from the Global South, where regulation is complicated by issues of development and claims on the world's resources. Indeed, 'There is a key misconception about China's manufacturing prowess. The Western Industrial Revolution began with technology innovation, whereas China's urbanization was mainly driven by global demand for manufactured products and thriving private businesses', as well as the vast movement of workers from the countryside to conurbations. Consequently, a future low-carbon transition in China would have far-reaching consequences for the world's ability to procure cheap plentiful numbers of objects.[58]

With 3D printing there is now the possibility of a convergence of supply and demand in a completely different reconfiguration of today's triad. An intimation of such a reconfiguration is 'reshoring': the return of manufacturing to post-industrial 'service' economies and the materialization of what are currently knowledge economies. Knowledge industries and sectors in the twenty-first century are increasingly aspiring to be a part of the material economy as well. For instance, Google is toying with the idea of entering the driverless car industry, in part due to the demand for time-space compression in commuting through synchronizing virtual mapping systems with physical infrastructures.[59] The problem for many knowledge economy companies is how to capitalize on the knowledge they produce in order to yield a profit.

3D printers are in a position to act as an interface between the material and knowledge economies in materializing intellectual capital in a decentralized fashion. As already shown, industrial designers make use of these technologies for making single instances of test models quickly and cheaply before their designs are sent to factories for bulk volume manufacturing. Increasingly 'rapid manufacturing' is becoming the norm for custom parts in limited instances of finished products – automobiles, aeroplanes and other products made in minimal batches.[60] There is then an expanding ecosystem where the technologies are used in a distributed manner: in small businesses and in the home for the personal production of mostly plastic prototypes.[61] The commonality in the spectrum of 3D printing is the ability to take a digitally created object, and using a given layer-by-layer building technique, recreate that object in a physical form. It is critical to note that this process typically involves not only access and understanding of infrastructure, materials supply and specialized software, but also significant knowledge in design (to model the engineering of the final structure, scaffolding and the printability correctly and effectively) and further finishing (as objects typically need post treatments or cleaning).

The materialization of data would seem to be the polar opposite of the focus of this special issue, namely the knowledge economy and universities. Not so, in fact universities – as producers, diffusers and preservers of new knowledge – are inching ever closer to the material economy due to the emergence in recent years of 3D printing within research centres, design schools, laboratories, and even academic libraries. Indeed, university libraries are the forerunners in the convergence between material and informational intellectual capital.[62] Invariably this innovation is in response to the demand from engineering and design students for rapid prototyping tools. However, many institutions are taking this on board in a similar fashion to the provision of centralized paper printing services by purchasing and making available 3D printers to all students, staff and researchers.[63] Others, such as Dalhousie University, are innovating across the 3D printing ecosystem by establishing online repositories of intellectual capital in the form of 3D model file collections.[64]

3D printing, known more formally as additive manufacturing, has had a renaissance in the last decade due to the commercialization of consumer level, mostly thermo-plastic extrusion, technologies and the consolidation of metal sintering in industry settings for end-user parts and products.[65] However, 3D printing's origins go back almost 40 years, when engineers were looking for better ways of prototyping designs in industry.[66] Since that era the field has matured and expanded to include an ever-growing suite of specialized equipment, printable materials and printing technologies.[67]

We do not propose that 3D printers are set to eat the knowledge economy from the inside out in a grassroots fashion, so to speak, even at the smallest scale platforms. Instead interconnectedness and reconfiguration is plausible, the topic of Chapter 6. We suggest entities in the knowledge economy will be mediators, suppliers and facilitators between 3D printing and consumers. In the process they will offset the risks involved with simply using these machines in the home environment without chaperoning. These risks include both intellectual property issues and personal safety ones, and both depend on centralized manufacturing assumptions.[68] In reorganizing themselves as triple helix collaborators and aligning the production of knowledge assets (inventions, patents, and designs) with industrial capacities of materialization, universities become hubs, partners or incubators in the convergent knowledge and material economy of the twenty-first century. While there remains paltry evidence of 'reshoring' – that is, bringing back mass production from 'emerging economies' – there is much evidence of job creation in the service sector and other niche manufacturing sectors, including production development, and in the low volume batch creation of complex, novel products.[69] Some suggest this poses the possibility of a 'manufacturing renaissance' in the US and other post-industrial nations alone. According to Autodesk's Chief Technology Officer, Jeff Kowalski:

> Manufacturing is probably going to be more localized than it has been. We won't be shipping as many raw materials around the world, producing things in lower-cost labor areas then sending it back. If manufacturing the actual production of something is effectively free, and more importantly, complexity is free, that can be performed locally.[70]

The potential for 3D printing in production has considerable efficiency opportunities for producers and consumers. Yet a key limitation of the 3D printing process for final version production, and not pre-production prototypes, is the high specific costs of the main current 'high end' contender technologies – that is, electron beam melting and direct metal laser sintering – in comparison to more conventional processes found on bulk factory assembly lines: injection moulding and machining.[71] It makes little economic sense to simply restock large-scale factories with clusters of 3D printers expecting the same or better cost advantages as the current system of production regardless of the efficiency gains in waste, quality control and labour. Moreover, many innovations in these three areas that apply to 3D printing, such as robotics, are in some cases applicable to machining and moulding, making these more cost effective too.

As we argue in this book 3D printing is as much a social phenomenon and is a core feature in rhetoric about the next phase of industrial revolution.[72] Estimations of this revolution understand it to be 'decentralized' since manufacturing no longer takes place in regional production clusters but in homes, offices, stores and, presumably, universities.[73] From our own viewing platform in the academy, a chief reason for tertiary institutions to transform into production and knowledge nodes within a decentralized industrial revolution is that they represent storehouses of intellectual capital with network links to local businesses, communities and citizens. Reshoring could allow universities to realize their 'third mission' of becoming wealth creators through materializing their intellectual capital.

## Shrinkage

In this section we examine the shrinking of supply chains and also appraise the idea that 3D printing and its associated ecosystem have the capacity to augment, shrink or even replace freight, supply chain capitalism and the complexities of distribution. Is the shrinkage of distance and time in freight and distribution a possible factor in 3D printing's ubiquity? Indeed it is already happening with pre-production through the ubiquity of 3D printing as rapid prototyping. Relevant too here is the push for leanness in distribution and the science of 'logistics'. The standardization of the shipping container and haulage infrastructure inter-modally in the mid-twentieth century was a precursor to globalization as well as to the deindustrialization of the US and other regions in the Global North creating both high levels of unemployment alongside a professionalized occupational structure.[74] The technology is a part of the logic of multi-modal integration that depends on uniformity of infrastructure and design conformity. Such controls operate within tightly managed, securitized spaces: ports and special economic zones. The growing costs of transportation, regional security, as well as questions about the future capacities of port infrastructures for growth, are examples of sources of 'friction'.

The forecasting of zero-length supply chains centres on both re-skilling, decentralizing and demobilizing manufacturing processes by placing distribution within the remit of the designers of products. In particular, the marked reduction in freight and print-to-order systems means more controlled production standards. Beyond production what will the implications of 3D printing be for freight? These occur across a number of dimensions: inventories, supply chains, custom products and ecological footprint.[75] From the logistics and supply chain management perspective there is anticipation in some quarters that 3D printing techniques auger new kinds of service operations involving a combination, rather than a substitution, of conventional forms of freight with digital manufacturing and equipment in use. For instance, the F-18 Hornet fighter plane supply chain was streamlined by producing parts closer to assembly lines and points of maintenance, however such innovation requires infrastructure investment to be of scale.[76]

Higher end products could also be 3D printed by companies on-site and on-demand using technically superior industrial printers and additive manufacturing technologies. Thus, in the not too distant future, only a small market of products would remain attractive for mass manufacturing. The ball is already rolling as middle-class cosmopolitan lives in the Global North already involve elements of a new system that is conducive to personal manufacturing: mass computer home-ownership; primarily services-oriented sectors such as information technology, design, architecture, education and governance; and cultural interests in customization, individuality and personal expression through fashion, aesthetics and craft. Moreover, there are development gains in the mass adoption of personal manufacturing wherein self-replication techniques, as demonstrated by the RepRap, mean that consumers could print their own 3D printer units at the cost of the raw materials. Potential consumers include the populations of China and India, already building services infrastructures, skills and resources in information technology.

Companies for a long time have been looking at the cost of freight and the possibilities of dramatically reducing costs by using reduced materials rather than finished commodities. Personal manufacturing offers a tipping point in how this might occur. A part of this trend is the potential of open source technologies and co- or peer- production to compete with for-profit systems where 'the individual is the coproducer of what he consumes'.[77] In fact this idea is now at the heart of the strategies of public and private firms that are no longer 'making things' and instead are central in knowledge hubs.[78] Notional 'desktop factories' would enable those unable to purchase objects produced elsewhere to produce them at low cost at the press of a button, with the potential to rebalance the global knowledge economy through printing and consuming, rather than trading, the intellectual capital.[79]

It is not only corporate interests that are imagining the mass 3D printing of objects, but not-for-profit ones too, who see a benefit from decoupling manufacturing from the social inequality and sustainability downsides to transoceanic global production networks.[80] The development of the not-for-profit open source 3D printer, chiefly the RepRap and its many for-profit progeny (including the US start-up company Makerbot), stems from a 'hobbyist ethic' for 'future craft'.[81] Progress here would mirror other large-scale forms of 'sharing' phenomena: car-pooling or distributed computing or software development, including Linux, Wikipedia and the Apache web-server standard.[82] These are amenable to global social movements emulating grassroots innovation in order to enact social change and reverse neoliberal market ideologies.[83]

The efficiency gains from dramatically lower transportation costs, as well as greatly more efficient additive manufacturing standards, mean that personal manufacturing can consume fewer resources than current systems. Indeed, if closed-cycles of feedstock can be innovated, as domestic 3D printer use increases on a par with 2D printers, new ways of procuring feedstock might be imagined along with new consumer practices of mass customization, which many companies already adopt in coming to terms with the 'age of access'.[84]

Organizations would not need the global logistics and business models that current MNCs deploy. One alternative to this would be the 'community enterprise', where localized cooperatives exist to provide public goods without central planning or control.[85] Such headless and/or bodiless organizations would act as communities rather than corporations, but would encounter resistance from existing powerful corporations.

perhaps more than many commentators on the impacts of e-tailing allow for. In the print book retail market there is evidence that people who buy e-readers purchase more books and read more. Moreover, consumers are able to publish their own books online digitally through repositories, only shifting to paper copies once having sold a certain number.[102]

Another reason 3D printing is implemented in potential social change around retail and consumption is due to the kinds of cultures in association with the technology. The spread of commodity cultures works through the marketing of objects to consumers.[103] Commodity cultures cater to modern identities condensing in the act of repeatedly purchasing products as a form of culture that is different from but not incompatible with 'high' forms of culture: religion, art, music and so on.[104] Commodity cultures cause friction, ranging from labour exploitation to issues of waste and concerns about domination (or the Coca-colonization) of other cultures.[105] 3D printing offers users a tool to resist or delay planned obsolescence and cyclic commodity cultures. Digital fabrication at a grassroots level is eminently suitable for putting 'commons-based, peer production' into practice.[106] In many cases these are in direct opposition to mass material cultures and transgress or resist short life-cycle products and planned obsolescence.[107]

Finally, there is the issue of simplicity and the burden of choice: 'When a customer is exposed to a myriad of choices, the cost of evaluating those options can easily outweigh the additional benefit from having so many alternatives'.[108] Certainly, many users of online systems report 'information overload' in comparison to simply browsing and selecting stock in a physical store environment. One response to the burden of choice in e-tailing is to trial virtual reality technology to give the illusion of physically 'being there' in a store of exhaustible objects, to try on items or to interact with store-people and the space. Research shows the balance of spatial visualization and sense of presence will be challenging to replicate sufficiently, given the variability in consumer demographics, gender being a crucial one.[109]

Mass customization is also conceivable in terms of a backlash against global consumer cultures. Decentralization, particularly of the means of production, is also compatible with counter-movements to the status quo.[110] For instance, leading up to India's independence from the British Empire the leader of the protest movement, Mahatma Gandhi, championed the spread of household cotton-spinning wheel (charkha) technologies to upset the British hegemony over cotton and textile production. The *Swadeshi* – a term from Sanskrit meaning 'made in one's own country' – movement relied on the popularization of incredibly simple clothing, the *khadi*, which anyone could spin themselves regardless of their wealth or level of expertise. Famously, Gandhi himself manufactured his own *khadi* and wore it on ceremonial visits overseas, and as a result the UK Prime Minister of the time, Winston Churchill, disparagingly referred to him as a' half-naked fakir' to much criticism.[111] It is not inconceivable that social movements espousing the rejection of variety and moderation might use the ubiquity of 3D printers in combination with freely shared digital templates for clothing, household goods and other functional items at the expense of cyclic consumer markets.

## Conclusion

As we have already seen, the shift towards leaner, just-in-time (JIT) production, distribution and consumption has been imminent since the late twentieth century. As we have found, early forms of 3D printing – that is, rapid prototyping – have their roots in product development. We have thus examined a range of possible technologies and practices. The issue is what kinds of system might get to be engendered here to bring the various elements together, so forming a new system. In this chapter the nascent presence of 3D printing within the existing system of production was examined in order to provide a scoping survey of a possible transition to ubiquity. The introduction of 3D printing into the current manufacturing system already involves many cost savings and efficiency gains that herald further progress in distributed, on-demand forms of 'lean' manufacture. Future areas of growth include customizing objects for consumers' particular interests, printing on demand, savings on raw materials – since much less gets thrown away in additive manufacturing – and the local adaptation of design elements so as to suit particular environments or sets of customers. In the next chapter we consider missing elements in this triad more closely: extraction and destruction where 3D printing is also relevant.

## Notes

1  A. Borgmann, 'The Force of Wilderness within the Ubiquity of Cyberspace'. *AI and Society* (2015): 1–5, doi:10.1007/s00146-015-0608-5

2  S. Turkle, *Life on the Screen: Identity in the Age of the Internet*. New York: Simon & Schuster, 1997.

3  T.J. Gerpott, 'Sms Use Intensity Changes in the Age of Ubiquitous Mobile Internet Access – a Two-Level Investigation of Residential Mobile Communications Customers in Germany'. *Telematics and Informatics* 32, no. 4 (2015): 809–22, doi:10.1016/j.tele.2015.03.005

4  H. Lipson and M. Kurman, 'Factory@Home: The Emerging Economy of Personal Fabrication Overview and Recommendations'. Washington, DC: House US Office of Science and Technology Policy, 2010. Occasional Papers in Science and Technology Policy 5.

5  C. Schubert, M.C. van Langeveld and L.A. Donoso, 'Innovations in 3D Printing: A 3D Overview from Optics to Organs'. *British Journal of Ophthalmology* 98, no. 2 (2014): 159–61, doi:10.1136/bjophthalmol-2013-304446. 160

6  L. Sandhana, 'The Printed Future of Christmas Dinner', *BBC*, 2010. Accessed 20 August 2012, http://www.bbc.co.uk/news/technology-12069495

7  D. Graham-Rowe, '"Gadget Printer" Promises Industrial Revolution', 2003. Accessed 24 August 2012, http://www.newscientist.com/article/dn3238-gadget-printer-promises-industrial-revolution.html?full=true&print=true

8  J. Wolfe, '3D Printing, Shapeways, and the Future of Personal Products', *Forbes*, 2012. Accessed 19 July 2012, http://www.forbes.com/sites/joshwolfe/2012/06/19/3d-printing

9  G.F. Davis, *Re-Imagining the Corporation, Real Utopias*. Ann Arbor: The University of Michigan, 2012. p. 2.

10  —, 'After the Corporation'. *Politics & Society* 41, no. 2 (2013): 283–308.

11  J. Tien, 'The Next Industrial Revolution: Integrated Services and Goods'. *Journal of Systems Science and Systems Engineering* (2012): 1–40. doi:10.1007/s11518-012-5194-1

12  C. Anderson, *The Long Tail: How Endless Choice is Creating Unlimited Demand*. London: Random House Business Books, 2012. p. 226.

13  C. Krogmann, *The Impact of Direct Digital Manufacturing on Supply Chains*. Berlin: GRIN Verlag, 2012.

14  M. Ratto and R. Ree, 'Materializing Information: 3D Printing and Social Change'. *First Monday* 17, no. 7 (2012). doi:10.5210/fm.v17i7.3968

15  M. Groenendyk and R. Gallant, '3D Printing and Scanning at the Dalhousie University Libraries: A Pilot Project'. *Library Hi Tech* 31, no. 1 (2013): 34–41. doi:10.1108/07378831311303912

16  J. Griffey, 'Absolutely Fab-ulous'. *Library Technology Reports* 48, no. 3 (2012): 21–4. p. 23.

17  S.R. Gonzalez and D.B. Bennett, 'Planning and Implementing a 3D Printing Service in an Academic Library'. *Issues in Science and Technology Librarianship*, no. 78 (2014). doi:10.5062/F4M043CC

18  J. Urry, *Offshoring*. Cambridge: Polity, 2014.

19  A.S. Blinder, 'Offshoring: The Next Industrial Revolution?'. *Foreign Affairs*, 1 March 2006. no pagination. Accessed 16 May 2012. http://www.foreignaffairs.com/articles/61514/alan-s-blinder/offshoring-the-next-industrial-revolution

20  J.E. McConnell, 'Geography of International Trade'. *Progress in Human Geography* 10, no. 4 (1986): 471–83. doi:10.1177/030913258601000401

21  N. Castree, A. Rogers and R. Kitchin, *A Dictionary of Human Geography*. Oxford: OUP, 2013. p. 72.

22  A.J. Scott and D.P. Angel, 'The Global Assembly-Operations of US Semiconductor Firms: A Geographical Analysis'. *Environment and Planning A* 20, no. 8 (1988): 1047–67.

23  S. French, A. Leyshon and N. Thrift, 'A Very Geographical Crisis: The Making and Breaking of the 2007–2008 Financial Crisis'. *Cambridge Journal of Regions, Economy and Society* 2, no. 2 (2009): 287–302, doi:10.1093/cjres/rsp013; A.L. Tsing, 'Empire's Salvage Heart: Why Diversity Matters in the Global Political Economy'. *Focaal – Journal of Global and Historical Anthropology* 64 (2012): 36–50, doi:10.3167fcl.2012.640104

24  D. Ernst and L. Kim, 'Global Production Networks, Knowledge Diffusion, and Local Capability Formation'. *Research Policy* 31, no. 8–9 (2002): 1417–29, doi:10.1016/S0048-7333(02)00072-0

25  N.M. Coe, P. Dicken and M. Hess, 'Global Production Networks: Realizing the Potential'. *Journal of Economic Geography* 8, no. 3 (2008): 271–95, doi:10.1093/jeg/lbn002; A. Smith, J. Pickles, M. Buček, R. Pástor and B. Begg, 'The Political Economy of Global Production Networks: Regional Industrial Change and Differential Upgrading in the East European Clothing Industry'. *Journal of Economic Geography* 14, no. 6 (2014): 1023–51, doi:10.1093/jeg/lbt039

26  A. Amin and A. Malmberg, 'Competing Structural and Institutional Influences on the Geography of Production in Europe'. *Environment and Planning A* 24, no. 3 (1992): 401–16; P. Filion, 'Potential and Limitations of Community Economic Development: Individual Initiative and Collective Action in a Post-Fordist Context'. *Environment and Planning A* 30, no. 6 (1998): 1101–23; C. Rowley, 'The British Pottery Industry: A Comment on a Case of Industrial Restructuring, Labour, and Locality'. *Environment and Planning A* 24, no. 11 (1992): 1645–50.

27  R. Martin, P. Sunley and J. Wills, 'Unions and the Politics of Deindustrialization: Some Comments on How Geography Complicates Class Analysis'. *Antipode* 26, no. 1 (1994): 59–76, doi:10.1111/j.1467-8330.1994.tb00231.x

28  C. Martin, 'Shipping Container Mobilities, Seamless Compatibility, and the Global Surface of Logistical Integration'. *Environment and Planning A* 45, no. 5 (2013): 1021–36.

29  J.J. Wang and D. Olivier, 'Port-Fez Bundles as Spaces of Global Articulation: The Case of Tianjin, China'. *Environment and Planning A* 38, no. 8 (2006): 1487–503.

30  A. Kellerman, 'Telecommunications and the Geography of Metropolitan Areas'. *Progress in Human Geography* 8, no. 2 (1984): 222–46, doi:10.1177/030913258400800203

31  N.M. Coe, 'Missing Links: Logistics, Governance and Upgrading in a Shifting Global Economy'. *Review of International Political Economy* 21, no. 1 (2014): 224–56, doi:10.1080/09692290.2013.766230

32  A. Kanngieser, 'Tracking and Tracing: Geographies of Logistical Governance and Labouring Bodies'. *Environment and Planning D: Society and Space* 31, no. 4 (2013): 594–610, doi:10.1068/d24611

33  A. Donovan, 'The Impact of Containerization: From Adam Smith to the 21st Century'. *Review of Business* 25, no. 3 (2004): 10–15. p. 13.

34  L. Crewe and M. Lowe, 'Gap on the Map? Towards a Geography of Consumption and Identity'. *Environment and Planning A* 27, no. 12 (1995): 1877–98, doi:10.1068/a271877

35  P.A. Jackson and B. Holbrook, 'Multiple Meanings: Shopping and the Cultural Politics of Identity'. *Environment and Planning A* 27, no. 12 (1995): 1913-30.

36  G. Slade. *Made to Break: Technology and Obsolescence in America*. Harvard: Harvard University Press, 2006.

37  J. Bulow, 'An Economic Theory of Planned Obsolescence'. *The Quarterly Journal of Economics* 101, no. 4 (1986): 729–50, doi:10.2307/1884176. 730

38  A. Herod, G. Pickren, A. Rainnie and S. McGrath-Champ, 'Waste, Commodity Fetishism and the Ongoingness of Economic Life'. *Area* 45, no. 3 (2013): 376–82, doi:10.1111/area.12022

39  J. Lepawsky and C. Mather, 'Checking in with Reality: A Response to Herod Et Al'. *Area* 45, no. 3 (2013): 383–85, doi:10.1111/area.12041

40  B. Warf, *Time-Space Compression: Historical Geographies*. Abingdon: Routledge, 2008.

41  D. Bak, 'Rapid Prototyping or Rapid Production? 3D Printing Processes Move Industry Towards the Latter'. *Assembly Automation* 23, no. 4 (2003): 340–5, doi:10.1108/01445150310501190. 341

42  B. Warf, *Time-Space Compression: Historical Geographies*. Abingdon: Routledge, 2008.

43  J. Peck and A. Tickell, 'Jungle Law Breaks Out: Neoliberalism and Global-Local Disorder'. *Area* 26, no. 4 (1994): 317–26, doi:10.2307/20003479

44  A. Smith, A. Stirling and F. Berkhout, 'The Governance of Sustainable Socio-Technical Transitions'. *Research Policy* 34, no. 10 (2005): 1491–510, doi:10.1016/j.respol.2005.07.005. 1493

45  C. Mota, 'The Rise of Personal Fabrication'. 2011 Proceedings of the 8th ACM conference on Creativity and cognition, Atlanta, Georgia, USA, ACM 2011 of Conference.

46  M. Ratto and R. Ree, 'Materializing Information: 3D Printing and Social Change'. *First Monday* 17, no. 7 (2012), doi:10.5210/fm.v17i7.3968

47  Y. Sugimori, K. Kusunoki, F. Cho and S. Uchikawa, 'Toyota Production System and Kanban System Materialization of Just-in-Time and Respect-for-Human System'. *International Journal of Production Research* 15, no. 6 (1977): 553–64, doi:10.1080/00207547708943149

48  A. Holl, R. Pardo and R. Rama, 'Just-in-Time Manufacturing Systems, Subcontracting and Geographic Proximity'. *Regional Studies* 44, no. 5 (2009): 519–33, doi:10.1080/00343400902821626

49  T.C.E. Cheng and S. Podolsky, *Just-in-Time Manufacturing: An Introduction*. London: Chapman and Hall, 1996.

50  A. Das and R.B. Handfield, 'Just-in-Time and Logistics in Global Sourcing: An Empirical Study'. *International Journal of Physical Distribution & Logistics Management* 27, no. 3/4 (1997): 244–59, doi:10.1108/09600039710170601. 248

51  M. Hesse and J.-P. Rodrigue, 'The Transport Geography of Logistics and Freight Distribution'. *Journal of Transport Geography* 12, no. 3 (2004): 171–84, doi:10.1016/j.jtrangeo.2003.12.004

52  A. Rai, R. Patnayakuni and N. Seth, 'Firm Performance Impacts of Digitally Enabled Supply Chain Integration Capabilities'. *MIS Quarterly* 30, no. 2 (2006): 225–46, doi:10.2307/25148729

53  K.E. Hill, 'Supply-Chain Dynamics, Environmental Issues, and Manufacturing Firms'. *Environment and Planning A* 29, no. 7 (1997): 1257–74.

54  B. Dankbaar, 'Lean Production: Denial, Confirmation or Extension of Sociotechnical Systems Design?'. *Human Relations* 50, no. 5 (1997): 567–83, doi:10.1177/001872679705000505. 567

55  B. Berman, '3-D Printing: The New Industrial Revolution'. *Business Horizons* 55, no. 2 (2012): 155–62, doi:10.1016/j.bushor.2011.11.003. 155
56  J. Urry, *Offshoring*. Cambridge: Polity, 2014.
57  P. Marsh, *The New Industrial Revolution: Consumers, Globalization and the End of Mass Production*. Padstow: Yale University Press, 2012. p. 61.
58  D. Tyfield, A. Ely and S. Geall, 'Low Carbon Innovation in China: From Overlooked Opportunities and Challenges to Transitions in Power Relations and Practices'. *Sustainable Development* 23, no. 4 (2015): 206–16, doi:10.1002/sd.1588. 214
59  M.N. Wexler, 'Re-Thinking Queue Culture: The Commodification of Thick Time'. *International Journal of Sociology and Social Policy* 35, no. 3/4 (2015): 165–81, doi:10.1108/IJSSP-06-2014-0048
60  N. Hopkinson, R.J.M. Hague and P.M. Dickens, 'Introduction to Rapid Manufacturing'. In *Rapid Manufacturing: An Industrial Revolution for the Digital Age*, edited by Hopkinson and Hague, 1-4. Chichester: John Wiley & Sons, 2006.
61  M. Michael, 'Process and Plasticity: Printing, Prototyping and the Process of Plastic'. In *Accumulation: The Material Politics of Plastic*, edited by Gabrys, Hawkins and Michael, 30–46. Abingdon: Routledge, 2013.
62  V.F. Scalfani and J. Sahib, 'A Model for Managing 3D Printing Services in Academic Libraries'. *Issues in Science and Technology Librarianship* 72 (2013), doi:10.5062/F4XS5SB9
63  S. Pryor, 'Implementing a 3D Printing Service in an Academic Library'. *Journal of Library Administration* 54, no. 1 (2014): 1–10, doi:10.1080/01930826.2014.893110
64  M. Groenendyk and R. Gallant, '3D Printing and Scanning at the Dalhousie University Libraries: A Pilot Project'. *Library Hi Tech* 31, no. 1 (2013): 34–41, doi:10.1108/07378831311303912
65  R.A. D'Aveni, 'The 3-D Printing Revolution'. *Harvard Business Review*. Harvard: Harvard Business Publishing, May 2015. Accessed 12 October 2015, https://hbr.org/2015/05/the-3-d-printing-revolution
66  K.V. Wong and A. Hernandez, 'A Review of Additive Manufacturing'. *ISRN Mechanical Engineering* (2012): 10, doi:10.5402/2012/208760
67  J. Moilanen and T. Vadén, '3D Printing Community and Emerging Practices of Peer Production'. *First Monday* 18, no. 8–5 (2013), doi:10.5210/fm.v18i8.4271
68  E.L. Neely, 'The Risks of Revolution: Ethical Dilemmas in 3D Printing from a US Perspective'. *Science and Engineering Ethics* (2015): 1–13, doi:10.1007/s11948-015-9707-4
69  B. Kianian, S. Tavassoli and T.C. Larsson, 'The Role of Additive Manufacturing Technology in Job Creation: An Exploratory Case Study of Suppliers of Additive Manufacturing in Sweden'. Type presented at the 12th Global Conference on Sustainable Manufacturing – Emerging Potentials, Malaysia, 2015.
70  R. Karlgaard, '3D Printing Will Revive American Manufacturing', *Forbes*, 2011. Accessed 3 August 2015,-. http://www.forbes.com/sites/richkarlgaard/2011/06/23/3d-printing-will-reviveamerican-manufacturing/
71  M. Baumers, P. Dickens, C. Tuck and R. Hague, 'The Cost of Additive Manufacturing: Machine Productivity, Economies of Scale and Technology-Push'. *Technological Forecasting and Social Change*, 102 (2016): 193–201, doi:10.1016/j.techfore.2015.02.015.193
72  B. Berman, '3-D Printing: The New Industrial Revolution'. *Business Horizons* 55, no. 2 (2012): 155–62, doi:10.1016/j.bushor.2011.11.003
73  S. Banerjee, '3D Printing: Are You Ready for the New Decentralized Industrial Revolution?', 2015 *Wired*. Accessed 19 August 2015. http://www.wired.com/insights/2015/02/3d-printing-decentralized-industrial-revolution/
74  O. Crankshaw and J. Borel-Saladin, 'Does Deindustrialisation Cause Social Polarisation in Global Cities?'. *Environment and Planning A* 46, no. 8 (2014): 1852–72.
75  J. Kietzmann, L. Pitt and P. Berthon, 'Disruptions, Decisions, and Destinations: Enter the Age of 3-D Printing and Additive Manufacturing'. *Business Horizons* 58, no. 2 (2015): 209–15, doi:10.1016/j.bushor.2014.11.005

76 J. Holmström and J. Partanen, 'Digital Manufacturing-Driven Transformations of Service Supply Chains for Complex Products'. *Supply Chain Management: An International Journal* 19, no. 4 (2014): 421–30, doi:10.1108/SCM-10-2013-0387

77 C. Marazzi, *The Violence of Financial Capitalism*. Los Angeles, CA: Semiotext(e), 2011. p. 51.

78 Ibid., p. 46.

79 A. Gorz, *Ecologica*. Kolkata: Seagull Books, 2010.

80 A. Smith, J. Pickles, M. Bu ek, R. Pástor and B. Begg, 'The Political Economy of Global Production Networks: Regional Industrial Change and Differential Upgrading in the East European Clothing Industry'. *Journal of Economic Geography* 14, no. 6 (2014): 1023–51, doi:10.1093/jeg/lbt039

81 L. Bonanni, A. Parkes and H. Ishii, 'Future Craft: How Digital Media is Transforming Product Design'. *Design* (2008): 2553–64, doi:10.1145/1358628.1358712

82 Y. Benkler and H. Nissenbaum, 'Commons-Based Peer Production and Virtue'. *Journal of Political Philosophy* 14, no. 4 (2006): 394–419, doi:10.1111/j.1467-9760.2006.00235.x

83 A. Smith and G. Seyfang, 'Constructing Grassroots Innovations for Sustainability'. *Global Environmental Change* 23, no. 5 (2013): 827–29, doi:10.1016/j.gloenvcha.2013.07.003

84 J. Rifkin, *The Age of Access: How the Shift from Ownership to Access is Transforming Modern Life*. London: Penguin, 2001.

85 D.R. Gress and R.V. Kalafsky, 'Geographies of Production in 3D: Theoretical and Research Implications Stemming from Additive Manufacturing'. *Geoforum* 60 (2015): 43–52, doi:10.1016/j.geoforum.2015.01.003

86 B.J. Cudahy, *Box Boats: How Container Ships Changed the World*. New York: Fordham University Press, 2006. p. 236.

87 S.J. Ramos, 'Planning for Competitive Port Expansion on the U.S. Eastern Seaboard: The Case of the Savannah Harbor Expansion Project'. *Journal of Transport Geography* 36 (2014): 32–41, doi:10.1016/j.jtrangeo.2014.02.007

88 K. Kelly, *What Technology Wants*. New York: Penguin Group USA, 2010.

89 D. Tyfield, 'Putting the Power in "Socio-Technical Regimes" – E-Mobility Transition in China as Political Process'. *Mobilities* 9, no. 4 (2014): 585–603, doi:10.1080/17450101. 2014.961262

90 G. Da Silveira, D. Borenstein and F.S. Fogliatto, 'Mass Customization: Literature Review and Research Directions'. *International Journal of Production Economics* 72, no. 1 (2001): 1–13, doi:10.1016/S0925-5273(00)00079-7

91 P.K. Wong, 'Creation of a Regional Hub for Flexible Production: The Case of the Hard Disk Drive Industry in Singapore'. *Industry and Innovation* 4, no. 2 (1997): 203–5.

92 V. Howard, *From Main Street to Mall: The Rise and Fall of the American Department Store*. Philadelphia: University of Pennsylvania Press, 2015.

93 P.L. Mokhtarian and I. Salomon, 'Modeling the Choice of Telecommuting: Setting the Context'. *Environment and Planning A* 26, no. 5 (1994): 749–66.

94 O. Rotem-Mindali and I. Salomon, 'Modeling Consumers' Purchase and Delivery Choices in the Face of the Information Age'. *Environment and Planning B: Planning and Design* 36, no. 2 (2009): 245–61, doi:10.1068/b34013t. 257

95 A. Kumar and Y.-K. Kim, 'The Store-as-a-Brand Strategy: The Effect of Store Environment on Customer Responses'. *Journal of Retailing and Consumer Services* 21, no. 5 (2014): 685–95, doi:10.1016/j.jretconser.2014.04.008

96 B. Smart, *Consumer Society: Critical Issues and Environmental Consequences*. London: SAGE Publications.

97 B. Bereitschaft and R. Cammack, 'Neighborhood Diversity and the Creative Class in Chicago'. *Applied Geography* 63 (2015): 166–83, doi:10.1016/j.apgeog.2015.06.020

98 J. Michael, 'It's Really Not Hip to Be a Hipster: Negotiating Trends and Authenticity in the Cultural Field'. *Journal of Consumer Culture* 15, no. 2 (2015): 163–82, doi:10.1177/1469540513493206

99 Z. Li, Q. Lu and M. Talebian, 'Online Versus Bricks-and-Mortar Retailing: A Comparison of Price, Assortment and Delivery Time'. *International Journal of Production Research* 53, no. 13 (2014): 3823–35, doi:10.1080/00207543.2014.973074. 3823

100  Ibid.
101  B. Fuchs, T. Ritz and H. Stykow, 'Enhancing the Blended Shopping Concept with Additive Manufacturing Technologies: Added Value for Customers, Retailers and Additive Manufacturers'. Type presented at the WEBIST 2012 – Proceedings of the 8th International Conference on Web Information Systems and Technologies, 2012.
102  E. House, 'Challenges Facing the UK Book Industry'. *Publishing Research Quarterly* 29, no. 3 (2013): 211–19, doi:10.1007/s12109-013-9320-9
103  L. Crewe and M. Lowe, 'Gap on the Map? Towards a Geography of Consumption and Identity'. *Environment and Planning A* 27, no. 12 (1995): 1877–98, doi:10.1068/a271877
104  P.A. Jackson and B. Holbrook, 'Multiple Meanings: Shopping and the Cultural Politics of Identity'. *Environment and Planning A* 27, no. 12 (1995): 1913–30.
105  U. Hannerz, *Cultural Complexity: Studies in the Social Organization of Meaning*. Columbia: Columbia University Press, 1992.
106  A. Smith and G. Seyfang, 'Constructing Grassroots Innovations for Sustainability'. *Global Environmental Change* 23, no. 5 (2013): 827–29, doi:10.1016/j.gloenvcha.2013.07.003
107  S. Hielscher and A. Smith, 'Community-Based Digital Fabrication Workshops: A Review of the Research Literature'. Sussex: SPRU Working Paper Series Science and Technology Policy Research University of Sussex. 8 May 2014. Accessed 13 October 2015, https://ideas.repec.org/p/sru/ssewps/2014-08.html
108  F.T. Piller and P. Blazek, 'Core Capabilities of Sustainable Mass Customization'. In *Knowledge-Based Configuration*, edited by Felfernig, Hotz, Bagley and Tiihonen, 107–20. Boston, MA: Morgan Kaufmann, 2014. p.113.
109  S.-Y. Yoon, Y.J. Choi and H. Oh, 'User Attributes in Processing 3D Vr-Enabled Showroom: Gender, Visual Cognitive Styles, and the Sense of Presence'. *International Journal of Human-Computer Studies* 82 (2015): 1–10, doi:10.1016/j.ijhcs.2015.04.002
110  T. Birtchnell, 'Elites, Elements and Events: Practice Theory and Scale'. *Journal of Transport Geography* 24 (2012): 497–502, doi:10.1016/j.jtrangeo.2012.01.020
111  S. Harcave, 'Gandhi's "Confession"'. *The Antioch Review* 8, no. 4 (1948): 507–9, doi:10.2307/4609307. 508

# 4
# RESOURCES

## Introduction

About 2 per cent of global oil is used to make a wide variety of manufactured goods in a rainbow of different polymers and as much as 95 per cent of packaging and bottling worldwide is derived from oil products. 3D printing in many low-end machines relies on ABS, which is a polymer derived from oil, or in some cases corn ethanol or PLA, which generally requires oil-based fertilizers anyway. High-end machines, which integrate lasers and even electron-beams, are able to experiment with many exotic materials – the company Objet has a patented rubber-like material called 'Tango', and the mail-order 3D printing company Shapeways provides users who upload designs with sandstone, glass, ceramics and even gold. In one novel process an open source robotic arm even 3D prints using soil.

In this chapter we focus on the stage before the production, distribution and consumption triad: resource extraction and the material economy. 3D printing is not simply a magic bullet allowing consumers to make as many objects as their hearts' desire. The fear that the world could be flooded with 3D printing failures, misprints and multiple copies of the same object with trivial variations masks a deeper issue. If consumers are able to easefully produce their own objects at the press of a button will there be a greater demand for resources in the form of powders, plastic filament wire and other feedstock on their own or contained within plastic and metal printer cartridges? In this chapter we address this concern through first appraising the era of the 'Plasticene': the sheer volume of plastic that is now entering the environment through the spread of our plastic-ridden society. In the second section we examine how much of the hype around 3D printing is in fact 'resource triumphalism' and we argue for a critical approach to the use of resources in digital fabrication. In the third section the notion of 'circular economies' is reviewed, namely the idea of sustainable systems of material use. In the fourth section we consider the 'decoupling' of resources from manufacturing entirely through utilizing recycled materials and innovative approaches to repair, replacement and design. In the fifth section we look to social practices as a source for reconfiguration of the resource systems that feed into the triad.

## The plasticene era

To gloss over the fact that 3D printing involves various materials whose origins lie in the extraction, transportation and conversion of resources at great cost to the environment would be disingenuous. Coal, natural gas and petroleum – that is, fossil fuels – account for the lion's share of the world's current use of energy. Notwithstanding the input of resources, namely the energy required to manufacture and transport objects, resources are the very fabric of the objects themselves. An estimated 4 per cent of world oil and gas production is used as feedstock for plastics and a further 3 to 4 per cent is expended to provide energy for their manufacture.[1]

Surprisingly, given this overwhelming reliance on fossil fuels, the mass production of plastic only began within living memory, in the 1950s, and is now ubiquitous to the point that it is pervasive in many marine and terrestrial environments where it was not intended for accumulation or use.[2] We are now in a 'plastic age' where globally in excess of 260 million tonnes are produced per annum, accounting for approximately 8 per cent of world oil production and where many different uses are put to this product, ranging from packaging to parts of transportation vehicles and all manner of consumer items.[3] We are literally swimming in a sea of plastic.

The technique of converting fossil fuels into synthetic materials (known in the industry as 'cracking') was one of the most game-changing innovations of the mid-twentieth century. The slogan of the Dupont corporation, first adopted in 1935 and only dropped in 1999, illustrates the perception of triumph over the natural world that human-made, or 'synthetic', polymers represented 'better things for better living through chemistry'.[4] These 'better things' were multitudes of novel, never-before-seen objects, and ones that replaced older ones and did not necessarily need to be plastic.

Prior to the plastics revolution couples receiving objects as wedding gifts could expect these tools to last their whole lives. Major brand 'Sunbeam' toasters from the 1950s made of metal, for instance, have suffered no depreciation in price because of their long lifespans in comparison to their plastic counterparts, which require routine replacement. In some cases those objects rendered obsolete were made from resilient, longer-lasting materials. With plastic becoming the material of choice, the use of others beneficial to the consumer in terms of longevity, but not initial outlay of cost, are now out of fashion: leather, metal, wood, and so on.

Beyond being disposable, and hence profitable for companies, plastics offered a gratifying aesthetic dimension to consumers, with form outdoing function and easeful disposability being marketed as a desirable feature. If it breaks simply purchase another one to replace it. A key innovation in plastic was that disposability principally became available once objects suffered any sort of visible decay regardless of functional impairment. Imperfections in the 'form' opposed to 'function' of an object, occurring either in the manufacturing process or in use – that is, scratches, scuffs or dents – were mostly tolerated or repaired prior to the Plasticene. Due to the low cost, unique material weaknesses (shattering in the same manner as glass or ceramics) and limited lifespans for plastic objects, replacement became a social norm.

Cultural historian Jeffrey L. Meikle articulates the shift in how people understood objects in terms of a pursuit of perfection linked to cultures of modernity, which plastic enabled through the cyclic replacement of objects and mass production using industrial injection moulding techniques. Disposability became associated with convenience and hygiene. Plastic environments became understood to be more healthy and satisfying of presumed ideas of future living.[5]

The mass disposal of plastic objects due to rapid decay in form or function, or both, and disposal in the environment are resulting in the present era imprinting itself in the strata. Reflecting the overwhelming use of plastic regardless of its fit-for-function and cosmetic vulnerabilities there is now much waste. Of the 300 million tonnes of plastics produced annually, about a third is thrown away soon after use, with a consequent impact upon the geological record: plastic in landfill of this magnitude will become a visible feature of future stratification. Thus commentators are now terming this possible fossil evidence for human consumption of plastic as the age of the 'Plasticene'.[6]

The presence of plastic in the geological record is concomitant with other features of the 'Anthropocene': rising emissions of GHGs, human population growth and biosphere degradation.[7] Scientific support for formal acknowledgement of the end of the Holocene due to a 'Great Acceleration' of global environmental crises represents a change in the intellectual climate, with ramifications for policymaking around resource-use, and plastic production, in the short term and long term.[8]

Certainly plastics are not readily locatable in the non-human world outside of volcanic landscapes where intense heat and pressure are able to melt hydrocarbons and create a chemical reaction. Plastics, while human-made, are essentially hydrocarbons with organic origins, hence the 'fossil' in the term 'fossil fuels'. For sociologist Adrian MacKenzie plastic is a hydrocarbon derivative bound up with and criss-crossing human bodies, landscapes, infrastructures and ecosystems – a result of biomass being processed.[9] In this optic humans and plastics are both akin as biochemical processes relating as metabolic relationships of living and non-living forms, which 'possess' resources.[10] Conceived of in this way, the problem is not necessarily that plastic is an artificial, toxic addition to the environment. Rather its pulse of use within human history reflects a lack of awareness of the fragility of the climate as one that is hospitable to people in comparison to catastrophes in other eras, such as the one in which fossil fuels such as petroleum were made a geological feature.[11]

Beyond the form of objects themselves is the form of the human body, its own decay, and the intimate association of plastic with people's identities and capacities for reinventing themselves physically and culturally.[12] Just as people in the twentieth century came to adopt cultures of disposability through the desire for perfection in the forms of objects, so too did they seek to enact the same cultures in themselves. What sociologist Anthony Elliott terms the idea of the 'new individual' comes about as a result of the plastics revolution. Synthetic materials, including plastics, are important in signs of status, in fashion as textiles and accessories, and in physical transformations, chiefly in cosmetic surgery.

If this is indeed the Plasticene Age then 3D printing is surely an extension of consumers' desire to reinvent their objects and their very selves through the medium of plastic and other 'controllable' materials. 3D printers, as introduced in the previous chapter, are unlikely to simply substitute for elements of the current mass production system: the economies of scale do not make sense. Instead, 3D printers will reach ubiquity if and when there is a social transformation, most likely towards a shift in custody of the means of production. The consequence for the environment of consumers-cum-citizens being granted the means of production would be momentous. While it would be thought that the democratization of manufacturing is an empowering trend, there is the scope for people to in fact produce more rather than fewer objects in bulk.

## Resource triumphalism

Next we aim to understand whether 3D printing represents a potential catastrophe of even greater plastic production, consumption and waste, or whether it could be a feature in the decoupling of resource-use from long-distance transportation and energy-inefficient manufacturing systems. Consumption (apart from production and distribution) is crucial here as forces of marketing coerce – softly or otherwise – consumer-citizens into short-term cycles of object replacement and obsolescence.

A way to move forward is to understand the 3D printer as just one in a number of icons – that is, mobile phones, automobiles, and personal computers – in one or multiple post-industrial, post-scarcity discourses. These motifs are prone to a trend that geographer Gavin Bridge asserts ignores 'the extraction, processing, and spatial transfer of material substances drawn from nature', namely, resource triumphalism.[13]

Three storylines support this discourse according to Bridge. The first is 'ecomodernization': the restructuring of post-industrial societies away from energy and resource intensive industries to service and knowledge intensive ones. For 3D printing, this first storyline slots in comfortably, wherein there is a foretold convergence of the knowledge and service economies with the material one as a result of ubiquitous 3D printing. Here industry rhetoric is stentorian. Dan Johns of the Bloodhound Project is aiming to build a supersonic car that is designed not only to go faster than the speed of sound, but also reach over 1,000 miles per hour. In a video enthusing about the benefits of 'additive manufacturing' Johns describes how the process impacts upon nearly all stages of the industrial process: using less material than subtractive manufacturing, resulting in less freight due to decreases in packaging sizes, and finally less waste due to reductions in unwanted products and by extension inventories.[14] 'If you look all the way back to the mining stage we can see that 3D printing uses much less material and therefore has a much more positive impact upon the environment'.[15] So here digital fabrication allows the global economy to have its cake and eat it as resources are digitized into non-existence as knowledge entities with no bearing in the material economy.

The second is 'dematerialization': a rise in GDP represents dropping demand for raw materials. In terms of the second storyline 3D printers are a new facet of the digital economy, bringing forth from the virtual realm consumers' informational products and materializing these far more efficiently than the present complex. In short, resources are capitalized on more effectively. An example of this thread is one of six 'megatrends' the management book *Leadership 2030* outlines. Front of stage in the trend of 'Technological convergence' are 3D printers. According to this feature 3D printing offers an economic advantage as it 'does away with economies of scale' sating consumer demand for individualization and 'turning industrial production on its head'.[16] So here digital fabrication allows the global economy to have its cake and eat it as resources are digitized into non-existence as knowledge entities with no bearing in the material economy.

The third is 'commodity prices', wherein resource availability has grown over time despite growing demand. In this instance, 3D printers are understood as a progression for de- or post-industrial regions (the US, UK, EU) as the manufacturing regions of the Global South align with the Global North. As President Barack Obama highlights 3D printing is just the latest technology undergirding a meritocratic ideal of consolidation: 'I want to make sure that if you work hard in this country, if you've got a good idea, if you're willing to put in some sweat equity, that you can make it here in America and live out your American Dream'.[17] From this viewpoint 3D printing is a method for achieving geopolitical security for post-industrial regions and at the same time boosting employment by allowing citizens to maximize their innovative capacity.

According to Bridge it is the distance of the commodity-supply space that is the reason for resource triumphalism in post-industrial settings. Indeed, 3D printing intimates a complete severance from uncomfortable questions about resource extraction and commodity production, as centralized manufacturing is reshored in the form of distributed manufacturing, without the negative aspects currently making it profitable and attractive to MNCs. As Bridge summarizes, 'the relative wealth and apparent independence of post-industrial economies from primary commodity production raises questions about the continued utility and value of resource extraction, particularly when activities such as forestry and mining are perceived to impose excessive social and ecological costs'.[18]

3D printing intimates the merger of manufacturing, distribution and services into a single technology, which the consumer manages and utilizes. 3D printing affords users with qualms about the ethical or environmental aspects of international trade to either support the 'reshoring' of manufacturing or opt out of the current system altogether. Next we consider counter-discourses to resource triumphalism heralding 3D printers as oppositional to the conservation, capitalization or consolidation of resources. These take into account sustainability to varying degrees.

## Circular economies

The current system is by no means efficient in its reuse of plastics.[19] Research that follows the 'geographical lives of commodities' involves ethnographic research by geographer Alison Hulme and traces the flow of bargain store commodities from China to the UK and back again as waste.[20] She and other social scientists show the crucial role informal waste pickers and pedlars in China, Bangladesh and other receivers in the Global South play in literally foraging for 'virgin' plastics – that is, those not derived from previously recycled plastics and lesser quality, which are less durable for reuse.[21] The conversion of repurposed objects into new objects with limited lifespans is an inefficient example of a 'circular economy', involving much energy waste in transoceanic transportation and subtractive bulk volume manufacturing.

Unsystematic recycling does not only apply to disposable knickknacks from bargain stores. Many conventional, brand products are also a part of global commodity systems, which are also partly circular, and involve long trails as well as mixtures of recycling and virgin materials. The trail of commodities spans many regions and the consumers of these products are largely unaware of the complexities of the global production networks sociologist Caroline Knowles details in her ethnographic work on the flip-flop.[22]

The flip-flop, a simple plastic footwear product, derives from oil extracted in the Middle East, for example Kuwait, and shipped by tanker to South Korea where it is then 'cracked' and mixed with other chemicals into thermoplastic granules ready for further transportation. After that the material is shipped again to Fuzhou, China, where it is distributed to factories where workers roll the granules and additional chemicals into sheets and cut them into the required shapes. From China the flip-flops are then shipped in bulk to retail distributors around the world through the complex containerization system where the cargo is informally seized by pirates in North Africa and smuggled as contraband for informal sale in markets in the Global South or formally moved through the Suez Canal and imported by major retailers in the Global North. Flip-flops' lifespans are short and once the straps break or the soles wears thin they are again shipped back again to Korea or China for recycling into more products or ultimately relegated to landfills. Flip-flops are just one example of 'fast fashion': the trend where consumers replace products before they are no longer functional due to, in part, enticement from marketing forces.

In order to assuage the demand for fast fashion, suppliers such as the mega-brand Inditex (owner of 11 brands including the popular 'Zara') require 'fast' transport and logistics in order to take advantage of both regional differences in the cost of production – thereby continuing to use distant clusters of suppliers at the expense of local ones – and fast cycles of consumer purchasing. In order to achieve this apparently contradictory impulse mega-brands centralize all of their distribution in the markets they target and are utterly dependent on cheap energy costs, and increasingly in air freight for high-end products.[23]

The label on many mass made, fast fashion objects sold globally – that is, 'Made in China' – is evidence of the transoceanic shipping and bulk volume factory production that go into their origin in markets outside of Asia. However so-called 'follow the thing' research demonstrates there is more to such life stories than their tags.[24] When considering the resources in jeans, the cotton is extracted and treated in many different countries, from Ukraine to Australia, before being dyed and shipped to production centres. Many components of objects are also subcontracted to other countries where additional processes take place, often at cheaper cost.

There is no product stream more demonstrably un-circular than electronic waste, or e-waste – a category presumably now also including 3D printers, alongside computer monitors and parts, paper printers, mobile phones and many other popular devices. In many instances, once transported to countries including Bangladesh, they are recycled down to their basic plastic and metal components, invariably by hand. Some geographers suggest that a more lucid appraisal than beginnings and ends in commodity flows within global production networks is as boundaries and edges and the actions that order them.[25]

## Decoupling

Decarbonization and dematerialization are both strong themes in future sustainability initiatives. The notion of 'decoupling' involves the reduction of 'global material and energy use, and carbon emissions, with only minimal impacts on improvements in living standards'.[26] As a process able to offer significant progress in terms of meeting an array of sustainability targets alongside other 'eco-technologies', including smart meters, electric batteries, hydrogen fuel-cells, photovoltaic panels and so on, 3D printing appears as just one of a host of techno-fixes, many of which are projected in order to 'make sure nothing really changes'.[27] Yet beyond techno-optimistic accounts, more critical commentators are now realizing that without infrastructural reconfigurations, alongside singular technology trends, the reigning system of mass manufacturing is not simply going to go away:[28] 'Manufacturing has always been done at scale, and required significant investment in fixed factories and machinery; 3D printing may vastly reduce these, as well as saving on transport and logistics costs'.[29] Taking into account reductions in transport and logistics another way to think about 3D printing is as a form of 'distributed manufacturing', with consequences for environmental sustainability arising from the 'decoupling' of production, distribution and consumption infrastructures. Current factors that contribute to drivers of perceived problems with the current system then undergo severance and decline. Yet infrastructural transformation does not just happen overnight and takes sturdy commitments from policymakers and master planners, not to mention financial entities and capital.[30]

The Intergovernmental Panel on Climate Change (IPCC) now recognize that 3D printing could play a part in 'decoupling' energy from manufacturing through reducing the need for transport.[31] Here 3D printing enters the picture due to its potential to reduce energy demand from inefficiencies, particularly in centralized industrial manufacturing and consequent global distribution. In the former, additive means of production allow greater precision with less waste and more scope for repair and retrofitting. In the latter, supply chains can be digitized or made leaner due to the mass distribution of the means of production, diminishing warehousing, and the shipping of empty space. Efforts to quantify potential reductions in emissions through 3D printing in industry to the year 2025 demonstrate reductions in the range of 130.5–525.5 mega tons of carbon dioxide, although forecasters admit it is still early days in terms of the technology's maturity.[32]

In this context the term 'distributed production' involves a blurring of consumption and production 'away from conventional mass production, with its long, linear supply chains, economies of scale and centralizing tendencies'.[33] In an exhaustive integrated review of a variety of disciplinary sources (chiefly engineering and production planning professions from a global sample of universities and research institutes) with the optic of sustainability three categories arose of note conveying consideration of a move away from mass manufacturing processes and systems.

First, the 'additive manufacturing process': rapid prototyping at the industrial scale reveals efficiencies in energy use, recycling and environmental impacts. Examples here include possible process energy measurements, life cycle inventories and experiments with recycling powder residues, for instance, of glass.

Second, mass customization and personalization are particularly overt in the product design sectors of fashion and textiles. Here, personalization in combination with life-cycle thinking and 'end-of-life strategies' is directed towards product longevity via uniqueness: users are so satisfied with objects of desire and art they disrupt the cyclic disposal and replacement extant in mass manufacturing today.

Third, there is a burst of interest in 'fabbing': open source systems that involve peer-to-peer forms of collaboration with a consequence for logistics. Fabrication occurs here in spaces equipped with small-scale digital manufacturing equipment, which individuals operate themselves. While not restrictive to additive manufacturing, so-called 'fablabs' foresee a noteworthy intervention of consumers into the process of manufacturing.

What the integrated review exercise shows is that product longevity; closed material loops; and more local community, as well as bespoke forms of manufacturing, are all topics of interest for vocational disciplines involved in planning for and achieving environmental sustainability.[34] 3D printing is just one part of this concept of 'distributed production', however all three categories above are relevant in forecasting the systemic affects of its ubiquity at different scales. In this concept, a combination of 3D printing with innovative design software could introduce a new kind of product with implications for sustainability.[35] So-called 'direct digital manufacturing' imagines 'batch sizes of one' in combination with 'online skill acquisition' and 'personalized high quality objects' in order to achieve lower energy use in production levels.[36] Such models foresee combinations of existing innovations – that is, the Internet and networked repositories – with future progress in 3D printing, particularly in metals, in order for a systemic transition to take place that is of the scale of a paradigm shift.

It is not just infrastructures of production, distribution and consumption that need replacement, substantial overhaul, or subtle retrofitting in order for 3D printing to instil itself as a method for sustainable decoupling. There is also waste to manage once objects are thrown away.

Aside from metals, plastics represent the greatest challenge for future societies facing the predicted increases in polymer-based products. The most widely available 3D printers, for instance the open source Reprap, use polylactic acid (PLA) and ABS. One possible option is so-called 'Recyclebots' working alongside 3D printers: household-scale semi-automated waste plastic extruders, recycling filament for use in open source 3D printers.[37]

## Social practices

Current supply systems have limited scope for circularity and are riskily reliant on comparative advantages and plentiful fossil fuels for corporate profit regardless of changes in the efficiency of the system. However, decoupling resources through 3D printing is not only possible through this technology substituting for transportation or for its possibilities for efficiency. There are also a number of imaginable ways that 3D printers could become a method for more localized sourcing. 3D printing could also drive a major shift in the pattern of consumption to services and products of higher value. Finally, 3D printing might have influence in the digitization of media and entertainment. In short, the social practices around object procurement must also be understood as a factor in resource use and dependency.

One area where the unique affordances of 3D printing are significant for social practices is in customers' ability to customize their own objects and hence value them more. Geographers Ruth Lane and Matt Watson argue that responsibilities for and relationships to products and materials are radically revisable through stewardship extending into the lifetime of objects, rather than applying once they are finished with. In short, such product stewardship at a domestic, household scale involves a change in material responsibility on the part of consumers.[38] An example here is the undercurrent of interest in and support for charity and second-hand stores. In these cases the age and obsolescence of objects are their attraction for consumers concerned with appearing individual and unique. The demand for aged, used, or vintage objects is one that 3D printing can facilitate for by offering services in reverse engineering or cloning unique products. In doing so it would be hoped that printed objects would also attract longer lifespans due to their unique properties divorced from fast fashion and planned obsolescence.

A signal demand for unique, custom objects is increasing in the now perceivable counter-cultural renaissance of small-scale making.[39] Rather than representing a step backwards in technological progress, these movements are indicative of, on the one hand, the capacity of corporations to be JIT, lean and flexible in response to the demand from consumers for variety; and on the other hand, the backlash against fast fashion, planned obsolescence and the profit-driven exploitation inherent in mainstream, bulk produced objects.[40] Indeed, the 'idea of manufacture as a discrete, stable and segmented component of the contemporary economy no longer stacks up against increased complexity, dependence on massive and sophisticated circuits of resource extraction, logistics management and financial capital'.[41]

One such counter-culture movement gone global is the online retail store 'Etsy', a launchpad for micro-entrepreneurial homeworkers, artisans, amateurs and small traders.[42] The food industry is similarly undergoing a shift to craft and local producers or 'foodies'. Once clusters of microentrepreneurs emerge out of local 'craft' cultures there is a global audience available to them, and economies of scale to a degree, as the microbrewery 'real ale' renaissance in the UK demonstrates. Parallels between the craft beer sector and 3D printing are particularly apt in regard to producers investing themselves in their products: 'the identity of the craft worker is seen as manifest in the product of their craft'.[43] 3D printing, in this framing, is a possible inroad for local cultures of making to locate new markets online or via social networking.

There is another dimension here for 3D printing in the democratization of craft expertise. Digital scanning, reverse engineering techniques and peer-to-peer file sharing enable those without technical skills to produce unique objects of perceived quality. 3D printers not only allow replicas to be made, they can also create hybrid re-assemblages, through the digital restoration of broken, decayed, historical or lost objects. These processes are not only useful for duplication or repair, but also offer new creative processes and aesthetic possibilities, for example an object inspired by the breakage of a beloved one.[44] Or an ancient Egyptian falcon's mummified skeleton, 3D printed from x-ray images.[45] However, not all would foresee positive effects from this potential for democratizing craft. The transition that 3D printing heralds potentially undermines the authenticity of cultural artefacts as well. The proliferation of replicas of original, priceless masterpieces with the ubiquity of 3D scanning and printing evinces and unsettles a widely held 'trust' in the value of objects.[46] For example, such a disruption to the economy of trust has ramifications for those who own, sell and esteem recherché objects – a multi-million, if not billion, dollar industry.

## Conclusion

Objects do not just appear out of thin air, they are made up of materials whose origins lie in the resource extraction of fossil fuels or to a lesser extent organic materials from agriculture. Currently, plastics require investments of energy and this consequently has an effect on the environment. In this chapter we have shown that the triad of production, distribution and consumption is heavily reliant on fossil fuels, and if 3D printing is to be a viable alternative then there will need to be a decoupling of energy from transportation and freight.

## Notes

1 J. Hopewell, R. Dvorak and E. Kosior, 'Plastics Recycling: Challenges and Opportunities'. *Philosophical Transactions of the Royal Society of London B: Biological Sciences* 364, no. 1526 (2009): 2115–26, doi:10.1098/rstb.2008.0311. 2115

2 D.K.A. Barnes, F. Galgani, R.C. Thompson and M. Barlaz, 'Accumulation and Fragmentation of Plastic Debris in Global Environments'. *Philosophical Transactions of the Royal Society of London B: Biological Sciences* 364, no. 1526 (2009): 1985–98, doi:10.1098/rstb.2008.0205

3 R.C. Thompson, S.H. Swan, C.J. Moore and F.S. vom Saal, 'Our Plastic Age'. *Philosophical Transactions of the Royal Society of London B: Biological Sciences* 364, no. 1526 (2009): 1973–76, doi:10.1098/rstb.2009.0054

4 S. Callahan, 'Dupont in April Dropped Its 64-Year Old Tagline, "Better Things for Better Living", in Favor of "the Miracles of Science"'. *B2B: Business Marketing*: Crain Communications, Inc., 1999. 1 June 1999. https://global-factiva-com.ezproxy.uow.edu.au/redir/default.aspx?P=sa&an=bzmk000020010825dv610006u&cat=a&ep=ASE

5 J.L. Meikle, *American Plastic: A Cultural History*. New Jersey: Rutgers University Press, 1997.

6 C. Reed, 'Dawn of the Plasticene Age'. *New Scientist* 225, no. 3006 (2015): 28–32, doi:10.1016/S0262-4079(15)60215-9

7 J. Urry, *Climate Change and Society*. Cambridge: Polity Press, 2011.

8 N. Castree and B. Christophers, 'Banking Spatially on the Future: Capital Switching, Infrastructure, and the Ecological Fix'. *Annals of the Association of American Geographers* 105, no. 2 (2015): 378–86, doi:10.1080/00045608.2014.985622

9 A. Mackenzie, 'Having an Anthropocene Body: Hydrocarbons, Biofuels and Metabolism'. *Body & Society* 20, no. 1 (2014): 3–30, doi:10.1177/1357034x13506470. 7

10 Ibid.

11 N. Clark, *Inhuman Nature: Sociable Life on a Dynamic Planet*. London: SAGE Publications, 2010.

12 A. Elliott, *Reinventions*, Shortcuts. Abingdon: Routledge, 2013.

13 G. Bridge, 'Resource Triumphalism: Postindustrial Narratives of Primary Commodity Production'. *Environment and Planning A* 33, no. 12 (2001): 2149–73, doi:10.1068/a33190

14 Bloodhound SSC, '3D Printing 1,000 Mph Car Parts', 2012 YouTube. Accessed 20 April 2012, https://youtu.be/0zwy8mX1UQ0

15 Ibid., no pagination.

16 G. Vielmetter and Y. Sell, *Leadership 2030: The Six Megatrends You Need to Understand to Lead Your Company into the Future*. New York: Hay Group Holdings, Inc., 2014. p. 132.

17 D. Hudson, 'The President Talks Manufacturing and Innovation at Techshop Pittsburgh', The White House, 2014. Accessed 15 September 2015, https://www.whitehouse.gov/blog/2014/06/18/president-speaks-manufacturing-and-innovation-techshop-pittsburgh

18 G. Bridge, 'Resource Triumphalism: Postindustrial Narratives of Primary Commodity Production'. *Environment and Planning A* 33, no. 12 (2001): 2149–73, doi:10.1068/a33190. 2168

19 N. Castree, 'Commodity Fetishism, Geographical Imaginations and Imaginative Geographies'. *Environment and Planning A* 33, no. 9 (2001): 1519–25, doi:10.1068/a3464

20 A. Hulme, *On the Commodity Trail: The Journey of a Bargain Store Product from East to West*. London: Bloomsbury Publishing, 2015.

21 N. Gregson, M. Crang, F. Ahamed, N. Akhter and R. Ferdous, 'Following Things of Rubbish Value: End-of-Life Ships, "Chock-Chocky" Furniture and the Bangladeshi Middle Class Consumer'. *Geoforum* 41, no. 6 (2010): 846–54, doi:10.1016/j.geoforum.2010.05.007. 847

22 C. Knowles, *Anthropology, Culture and Society: Flip-Flop: A Journey through Globalisation's Backroads*. London: Pluto Press, 2014.

23 A.I. Escalona Orcao and D.R. Pérez, 'Global Production Chains in the Fast Fashion Sector, Transports and Logistics: The Case of the Spanish Retailer Inditex'. *Investigaciones Geográficas, Boletín del Instituto de Geografía* 2014, no. 85 (2014): 113–27, doi:10.14350/rig.40002

24 A. Brooks, 'Systems of Provision: Fast Fashion and Jeans'. *Geoforum* 63 (2015): 36–9, doi:10.1016/j.geoforum.2015.05.018

25 J. Lepawsky and C. Mather, 'From Beginnings and Endings to Boundaries and Edges: Rethinking Circulation and Exchange through Electronic Waste'. *Area* 43, no. 3 (2011): 242–9, doi:10.1111/j.1475-4762.2011.01018.x

26 H. Schandl, S. Hatfield-Dodds, T. Wiedmann, A. Geschke, Y. Cai, J. West, D. Newth, T. Baynes, M. Lenzen and A. Owen, 'Decoupling Global Environmental Pressure and Economic Growth: Scenarios for Energy Use, Materials Use and Carbon Emissions'. *Journal of Cleaner Production* (2015), doi:10.1016/j.jclepro.2015.06.100. 1

27 E. Swyngedouw, 'Apocalypse Forever? Post-Political Populism and the Spectre of Climate Change'. *Theory, Culture & Society* 27, no. 2–3 (2010): 213–32, doi:10.1177/0263276409358728. 222

28 N. Castree and B. Christophers, 'Banking Spatially on the Future: Capital Switching, Infrastructure, and the Ecological Fix'. *Annals of the Association of American Geographers* 105, no. 2 (2015): 378–86, doi:10.1080/00045608.2014.985622

29 A. Sissons and S. Thompson, *Three Dimensional Policy: Why Britain Needs a Policy Framework for 3D*. London: Big Innovation Centre, 2012. p. 8.

30 N. Castree and B. Christophers, 'Banking Spatially on the Future: Capital Switching, Infrastructure, and the Ecological Fix'. *Annals of the Association of American Geographers* 105, no. 2 (2015): 378–86, doi:10.1080/00045608.2014.985622

31 R. Sims, R. Schaeffer, F. Creutzig, X. Cruz-Núñez, M. D'Agosto, D. Dimitriu, M.J. Figueroa Meza, L. Fulton, S. Kobayashi, O. Lah, A. McKinnon, P. Newman, M. Ouyang, J.J. Schauer, D. Sperling and G. Tiwari, 'Transport'. In *Climate Change 2014: Mitigation of Climate Change. Contribution of Working Group III to the Fifth Assessment Report of the Intergovernmental Panel on Climate Change*, edited by Edenhofer, Pichs-Madruga, Sokona, Farahani, Kadner, Seyboth, Adler, Baum, Brunner, Eickemeier, Kriemann, Savolainen, Schlömer, von Stechow, Zwickel and Minx, 599-670. Cambridge: Cambridge University Press, 2014.

32 M. Gebler, A.J.M. Schoot Uiterkamp and C. Visser, 'A Global Sustainability Perspective on 3D Printing Technologies'. *Energy Policy* 74 (2014): 158–67, doi:10.1016/j.enpol.2014.08.033

33 C. Kohtala, 'Addressing Sustainability in Research on Distributed Production: An Integrated Literature Review'. *Journal of Cleaner Production* 106 (2015): 654–68, doi:10.1016/j.jclepro.2014.09.039. 654

34 Ibid.

35 W.B. Arthur, *The Nature of Technology: What It is and How It Evolves*. New York: Free Press, 2009.

36 D. Chen, S. Heyer, S. Ibbotson, K. Salonitis, J.G. Steingrímsson and S. Thiede, 'Direct Digital Manufacturing: Definition, Evolution, and Sustainability Implications'. *Journal of Cleaner Production* 107 (2015): 615–25, doi:10.1016/j.jclepro.2015.05.009

37 C. Baechler, M. DeVuono and J. Pearce, 'Distributed Recycling of Waste Polymer into Reprap Feedstock'. *Rapid Prototyping Journal* 19, no. 2 (2013): 118–25, doi:10.1108/13552541311302978

38 R. Lane and M. Watson, 'Stewardship of Things: The Radical Potential of Product Stewardship for Re-Framing Responsibilities and Relationships to Products and Materials'. *Geoforum* 43, no. 6 (2012): 1254–65, doi:10.1016/j.geoforum.2012.03.012

39 C. Carr and C. Gibson, 'Geographies of Making: Rethinking Materials and Skills for Volatile Futures'. *Progress in Human Geography* (2015), doi:10.1177/0309132515578775

40 B. Smart, *Consumer Society: Critical Issues and Environmental Consequences*. London: SAGE Publications, 2010.

41  C. Carr and C. Gibson, 'Geographies of Making: Rethinking Materials and Skills for Volatile Futures'. *Progress in Human Geography* (2015), doi:10.1177/0309132515578775. 5

42  S. Luckman, 'Women's Micro-Entrepreneurial Homeworking'. *Australian Feminist Studies* 30, no. 84 (2015): 146–60, doi:10.1080/08164649.2015.1038117

43  T. Thurnell-Read, 'Craft, Tangibility and Affect at Work in the Microbrewery'. *Emotion, Space and Society* 13 (2014): 46–54, doi:10.1016/j.emospa.2014.03.001. 47

44  A. Zoran and L. Buechley, 'Hybrid Reassemblage: An Exploration of Craft, Digital Fabrication and Artifact Uniqueness'. *Leonardo* 46, no. 1 (2013): 5–10.

45  P. Anton Du, S. Ruhan, S. Liani Colette, E. Johan, J. B. Gerrie, I. Salima and C. Izak, 'Three-Dimensional Model of an Ancient Egyptian Falcon Mummy Skeleton'. *Rapid Prototyping Journal* 21, no. 4 (2015): 368-72, doi:10.1108/RPJ-09-2013-0089.

46  T.A. Easton, 'A Recession in the Economy Trust'. In *Values and Technology: Religion and Public Life*, edited by Ricci, 159–69. New Jersey: Transaction Publishers, 2011.

# 5

# TRANSITIONS

## Introduction

We have noted that it is estimated that the global 3D printing market will reach approximately US$3 billion by 2018.[1] Although there is a growing presence it only partially matches the hype around its possible impacts upon society. In this chapter we examine the speculation that 3D printing will 'change the world'.[2] What this now widespread commentary infers is that 3D printing is an element in what social scientists term a 'socio-technical transition'. There are important world-scale externalities driving a transition involving 3D printing: climate change and energy depletion. There are also drivers internal to the system: social inequality, infrastructural capacity and consumer sentiments. Lastly, beyond internal and external drivers of transition there are those arising from the nature of technological evolution itself.

As explained in the first chapter, 3D printing is not simply a more efficient, cost-effective and flexible form of factory automation: 'Most predictions are based on a technological analysis that compares the abilities of the new 3D desktop printers with the industrial capabilities of mass-production ... if we follow this argument, we would see the Internet as no more than a technical improvement on the telegraph'.[3] Rather than simply reaching ubiquity due to satisfying a specific niche in society, 3D printing is now a topical subject because of its intimations of revolution in society.[4] Despite the estimations of a pending revolution, there is much confusion in these domains about 3D printing's affordances and limitations: what exactly will it upheave? How will 3D printing disrupt some or all of the elements within the existing system? Not according to the economics of 3D printing as it stands: 'instead of looking at it as a substitute for existing manufacturing, we should look to new areas where it can exploit its unique capabilities to complement traditional manufacturing processes'.[5]

As the Managing Director of Australia's Objective3D – a commercial additive manufacturing 'factory' – admits, there is no capacity to replace high volume industries such as car making or food canning.[6] Even the Guinness World Record-holders, Airwolf 3D 'Wolf Pack', who operated the most 3D printers in one place simultaneously (159 individual units), pales in comparison to the assembly line factories in operation in parts of Asia today that involve thousands of machines and an equal number of labourers.[7] If it is unlikely that 3D printing will simply substitute for the current system, then another explanation is needed. What the commentaries on a new industrial revolution share in common is awareness of the object diversity 3D printing promises. Mass customization appeals to a global consumer culture dissatisfied with homogeneity. The argument here claims consumers crave novelty, variety and choice far beyond what mass manufacturing is capable of providing. By bringing the consumer closer to production, 3D printing promises a new intimacy with making things, one that has been lacking ever since the introduction of mass machine labour to craft industry in the early twenty-first century. Much of this demand appears to stem from digital innovations in traditional craft.[8] It could be that 3D printing will neither scale up in a centralized fashion nor replace mass manufacturing. 3D printing could indeed be 'a complement, a new tool in the box … both in making existing products better as well as being able to manufacture entirely new ones that we previously could not make'.[9] Beyond the technological capabilities and limitations that signal a transition are social ones. Dissatisfaction with consumer culture could manifest in support for an alternative system. A third possibility is 'demassification' as creative economies encourage a large-scale adoption of customization technologies.[10] The 'game-changing' features of 3D printing are not in the same paradigm as the mid-twentieth century imagination of the impacts of automation as a method of substituting for humans: anthropomorphic machines directly replacing physical labour, 'pushing' them out of factories. While the current system is far from homogeneous, instead of striving to satisfy consumer demand for choice, 3D printers offer a far more broad range of products that are tailored to individuals wants and needs.[11] There is then scope for 3D printing to hold transitional features akin to other social innovations.

## Massification

In this section we aim to clarify why 3D printing is a prominent example of a technology driving a forthcoming 'industrial revolution'. Other historical socio-technical transitions towards the massification of industry are informative for 3D printing. Industry commentators', journalists' and academics' predictions of imminent social change mirror the rise to ubiquity of personal computers in the 1980s.

As Chapter 1 laid out, one of the most crucial of the features 3D printing affords is product customization without craft expertise or intermediaries: 3D printers enable the manufacture of single instances of objects from digital files without requiring an individual investment of human expertise, beyond the initial pre-production design process. Such an affordance has gravity in the object-dense, yet reasonably uniform, batch-driven consumer cultures of the twenty-first century, at least in the materially rich Global North. Within such cultures people express and form their identities through purchasing, using and making symbolic capital derived from consumer goods produced within large, energy-intensive factories in a high-energy, high-mobility system.[12]

In both consumer and industrial settings within the current system experts estimate 3D printing to be emerging as ubiquitous in the same fashion as numerous 'small' technologies in use every day now. Personally worn 'wrist' watches became ubiquitous in the first half of the twentieth century – a trend from which the sociologist Georg Simmel derived his social theories on 'clock-time' as a mental way of life.[13] Automobiles reached ubiquity in the second half of what became the century of the car. In the late 1980s the managing director of the Nissan Science Foundation, Tomokazu Tokuda, projected that cars in the 1990s would be classed 'humanware' due to a 'total relationship with social activities and individual lifestyles'.[14] Personal computers, mobile phones and indeed electronic paper printers became ubiquitous within the lifetimes of those born in the twenty-first century: a digital era with a likeminded 'Generation Wired'. 3D printing is now projected to be close to emerging as a ubiquitous humanware in the same fashion by some commentators.

The current 'mass' manufacturing system evolved to be high-energy due to the remarkable discovery of vast reserves of petroleum, coal and natural gas 'fossil fuels' in the mid-twentieth century.[15] The mass manufacturing system is also one of high-mobility because the automated assembly line could not have remained profitable without the rapid upscaling of transoceanic freight utilizing logistics innovations in telecommunications alongside automation technology in ports and warehouses.[16]

All of the above technologies owe their ubiquity to the automation of factory-centred manufacturing in the twentieth century, chiefly the assembly line, and later injection moulding and CNC milling. These ubiquities came about due to tipping points pre-empting systemic transitions: the 'synching' of people's daily routines to clocks around 'universal' Greenwich Mean Time and the forceful mass adoption of private transport through a combination of lobbying by fossil fuel and automobile corporations; elite support in secure and luxurious mobility; petroleum extraction and exploitation; and policy and planning pushes to make cities commutable by car.[17]

The term 'automation' derives from the 1940s and refers to the introduction of automatic equipment in a manufacturing or other process or facility. It is no coincidence that it is at this time that many automatic devices enter the home, workplace and public space in many parts of the world: washing machines, kitchen utensils, escalators, and so on. Uniform objects made in bulk are the order of the day due to the natures of production and distribution that arose in the twentieth century in this second industrial revolution, notably through industrialists Henry Ford and Malcom McLean's assembly line and containerization innovations respectively. Despite speculation, there is much uncertainty about what 3D printers' features mean for objects made, moved and marketed within the systems of mass production, distribution and consumption. Chiefly, 3D printers' ability to automate to varying degrees the manufacture, freight and retail of objects – in some cases singly – suggests a capacity to upstage current processes.

Automation was an integral part of the social transformation that characterized the last few centuries, which saw machines emerge capable of replicating people's work practices. Now we know that people were not all made redundant by automation – machines did not substitute wholesale for human labour. In coming to terms with the impact of automation in recent times we will focus on centralization as well as substitution in order to reflect on how 3D printing might have a similar reverberation. Why did automation cause centralization in the first place? A core trade-off in the first Industrial Revolution – historically located in Britain in the mid- to late nineteenth century – was between the quality of handicraft manufacturing and the quantity of machine-made manufactures. Historians of this era are unclear about the role of technological innovation in social transformation. On the one hand, technical change due to inventive activity in the cotton and iron industries was indeed 'revolutionary'. Rather than simply replacing human labour many of the innovations increased productivity through speed and volume. By the late nineteenth century there was little room for handicrafts (weaving or gun-manufacture) in these sectors and the surge in output resulting in economic growth for Britain was by all accounts unprecedented. On the other hand, the role of purely technological innovations has been overblown. The steam engine appears to have had much less of an effect across the whole of Britain's economy.[18] Whatever the case, commentators are generally unanimous that the social impacts of the Industrial Revolution were the migration and centralization of people in cities, the conversion of the workforce from skilled cottage industries to unskilled factory work in centralized facilities, and the onset of significant demand for fossil fuels and consequent carbon emissions through centralized energy use.

Social science commentary on the impacts of centralization due to automation – invariably in factories – arose in the nineteenth century in studies of the importance of science and its techniques to human thought,[19] the psychological consequences of the mechanical division of labour,[20] and the motivation of exploitation and inequality in the workplace.[21] Centralization was understood to be momentous for human life as it changed where people lived (large urban centres), how they worked (by the clock) and their family and community relationships (separate work and private lives). As we will see, what makes 3D printing different from earlier waves of automation is that this process involves a trend towards decentralization rather than centralization.

Due to its scope for change 3D printing is being couched as a pending 'industrial revolution for the digital age'.[22] By some estimates 3D printing's ubiquity will mean a decentralization of production as momentous as the centralization of production that took place in the initial Industrial Revolution. The 'massification' of 3D printing in line with previous technologies would suggest a consequent substitution for factory-centred mass production, container-centred mass distribution, and warehouse-centred consumption. Consumers able to simply print their own objects unique to themselves are unlikely to demand mass manufactured ones at the same time, or so the logic goes. Centralization is important for 3D printing. Unlike the dominant types of automation found in factories around the global today, 3D printing intimates a trend towards decentralization: of bringing individuals into a further engagement with manufacturing processes.

There are three major themes that frame automation throughout the twentieth century which shed light for us on 3D printing's projected ubiquity, introduced already in Chapter 2. First there is the notion of substitution. In the middle of the twentieth century advanced machines were fathomed to be closely capable of not only replicating human labour but also rendering it obsolete. While some greeted this notion with glee and speculation about the boundless extent of newfound leisure time, others understood it as a threat to civilization, and an encroachment on wellbeing and livelihoods. The ubiquity of personal computers accompanied a further phase of concern as automation spread to brain as well as muscle work. Office-work was also imagined to be substitutable. The hypothesis goes that 3D printers are just another stage in the automation and replacement of human labour, this time in a decentralized manner. 3D printing is understood to be more inclusive and democratizing than assembly line-driven mass manufacturing. In a world with 3D printers present in the home, office, high street or library, there is the temptation to predict substitution, just like in the online retail revolution. It is necessary to reflect critically upon suppositions that technological innovations accompany the substitution of one technology for another.[23]

Second is the notion of homogenization as automation was observed to be ushering in consumer cultures of habitual spending and floods of identical objects on global markets. Mass manufacturing homogenizes culture through forcing consumers to desire and purchase uniform objects in a system designed to extract maximum efficiency and profit. 3D printers provide an alternative to the production–distribution–consumption triad in affording consumers the ability to mass customize.

Third, the massification of object production is the prevailing system today; however, there are other possibilities too and a popular one is the notion of demassification wherein 3D printing is ubiquitous, yet neither substitutive of, nor a product dependent upon, mass manufacturing. Demassification represents a plausible explanation for 3D printing's place in a new industrial revolution, beyond substitution or homogenization, that 3D printers will provide an alternative to the production–distribution–consumption triad in affording consumers the ability to mass customize objects themselves.

## Demassification

Beyond substitution and homogenization there is another third possibility for the explanation as to why 3D printing is being cast as the next industrial revolution, namely demassification. In 1979 futurist Alvin Toffler coined the term 'demassification' in order to denote an era where there would be either no demand for or no capacity to supply massified objects to global consumers. Toffler's focus was skewed towards the demand-side. He understood there to be a looming backlash in the massified industrial society of the twentieth century as a result of the establishment of global standards common across the world: these had sown the seeds of their own destruction. The consequence, so Toffler thought in the 1970s, was a looming generational transition – the so-called 'Third Wave' – that rejects global standards and the veneer of uniformity in modern life due to massification: the cultural domination of English; design homogenization; the impact upon family life and community of distant workplaces and suburban sprawl, a by-product of people's need to earn a living in order to afford access to massified systems. Toffler's awareness of forces fostering discontent towards homogenization in everyday life is now more than ever relevant and there are others that he did not flag. Toffler foresaw the future implications of computer control in manufacturing and even plastic extrusion wherein short to single runs of objects would become possible.[24] As consumer-level 3D printers now enter the marketplace – even found for sale in Walmart – many of Toffler's ideas about demassification are resurfacing.

'Massification' is a term that has come to define the rapid period of industrialization of the past century wherein industrial processes, standards and sweeping societal transformations enabled vast numbers of people globally to gain access to an unprecedented wealth of commodities and information: the era of modernity. The massification of many systems of supply in the twentieth century (energy, water, manufacturing and many others) went hand-in-hand with similarly massified 'social practices' that accompanied and reinforced systems of demand (marketing and retail).[25] Throughout the twentieth century there arose 'normal' social practices around certain ubiquitous technologies produced from key locations and systems in a centralized fashion. Examples include automobiles, mobile phones, white-goods, household amenities, product packaging, clothing and food and hygiene products. Before massified systems, certain of these 'things' would have been available to a majority of people, however the difference is that massification ironed out any inconsistencies (good or bad) that occurred at a local, individual level that made individual objects slightly or largely different in quality or aesthetic appearance.

The massification of manufacturing systems is a distinctly twentieth-century phenomenon for a handful of key reasons. The massification of humans made mass manufacturing profitable. A boon to the mass manufacturing system was the discovery of oil reserves on the order of billions of barrels a year that provided standardized, synthetic and predictable materials (invariably plastics) and plentiful energy supplies to fuel production. In addition, Americanization became a 'rooted cosmopolitanism' in the mid-twentieth century as a global cultural phenomenon (global America). America's culture became integrated, emulated and mutated in very many other cultures with some core familiar features: the nuclear family, democracy, suburban sprawl, malls, automobile fetishization and dependency, corporate idols (Disney, McDonalds), media idols (Marilyn Monroe, James Dean, Elvis Presley), and so on.[26] The twentieth century was an era of mass mobility and the massification of distribution through containerization moved objects around the world with pre-dictability and profit. Finally, all of these factors set in motion systems to cater to demand for a variety of what can be understood as 'high-carbon' social norms on a global scale.

It is not just the mass production of physical objects that the term applies to, but also of data. The massification of universities due to the advent of purely Internet-based systems – Massive Open Online Courses (MOOCs) – is a discussion point attracting much attention. Beyond its reference to the systems in place in produc-tion, distribution and consumption, the term 'massification' is meaningful in this book because of its dualistic meaning – of both positive and negative qualities – in wider use.

Something Marcuse did not foresee is the increasing choice that the automation of production, in the form of 3D printing, can bring. Consumers of MOOCs and mass manufactured objects share a sense of disconnection from the processes extant in these systems, and this is a chief reason for the pejorative application of massifica-tion. The online university, the assembly line factory and the industrial farm are sites that are unfamiliar, and hence disconnected, from those whose lives are nev-ertheless shaped by the products that emerge from them. The same goes for infor-mation proliferating through media systems about a world beyond the family or local community. The disconnection of consumers from massified systems has led to the term's frequent critical use in media studies. Hence massification also refers to 'homogenization' and a relative fin de siècle that heralds a coming phase shift to 'demassification'. The emergence of the prefix here is crucial for foresight on 3D printing for a suite of reasons discussed later in this book.

Therefore massification is an ever-present feature in the everyday lives of those in the high-carbon Global North and is a blessing and a curse for two principal reasons. The former because 'standards' of living are currently judged as never having been so high, and the latter because massification systems and products are unequivocally problematic for the climate-as-we-know-it, as well as being deleteri-ously homogeneous to social, physical and cultural differences. What will demas-sification be like and how will it gel with new technologies and social practices? Sociologist Susan Luckman and others show that there is motion in the current system towards demassification.[27]

A recent exercise on the future of the smart city in the US to the year 2037 with urban logistics executives working in the small business sector envisaged a city landscape with axes of uncertainty of low global trade and high resource availabil-ity. The vision of demassification, given the title 'Millions of Markets', understood a significant degree of self-reliance in energy agriculture, manufacturing alongside migration to smaller urban areas. The chief technology driver of this scenario and accompanying 'prototypes' is high-end 3D printing becoming ubiquitous at the local scale – in this case Boston – with high individual engagement in the processes of making. Inspirations for involvement of individuals include: comfort, conveni-ence, aesthetic qualities and personal utility. The authors conclude that 3D printing in this instance could 'foster community cohesion and resilient economic futures with more local communities self-sufficient in their economic production'.[28] A companion study to 2033 imagines a similar systemic transformation around a new method of production opening up an entirely new market in terms of customizable source code being traded instead of finalized products.[29]

## Reconfiguration

We have thus far established that transition in the production–distribution–consumption triad is being driven by multiple factors, ranging from global social inequalities, cultural diversity and economic growth, to climate change, energy security and environmental degradation. A future where 3D printing is ubiquitous and demassification is widespread will not resemble the interlocking systems at work today. There will be momentous change arising from a transition to a wholly different method for consumers to procure objects. Furthermore, there are factors policymakers and leaders need to take into account when participating in, leading, or shepherding change without social unrest and even collapse. Contrastive positions are extant today for those in governance to take into account and to inform the policymakers they look to for advice. Social scientist Frank Geels and colleagues have usefully articulated three of these main positions from research on sustainable production and consumption research, which current projections of socio-technical transition for the production–distribution–consumption triad tend to follow.[30] These analytical heuristics are not absolutist, but roughly align with different models of governance and real-world instances of change. They result in the different kinds of influence that policymakers and leaders will exercise over transitions.

First Geels and colleagues articulate a 'market' mode of governance driving a 'reformist' position to change at the social scale. The political and academic orthodoxy support this view of transition and incumbent firms and mainstream policymakers enact it. Here there is a presumption of business-as-usual and any transformations arise from efficiency, economic growth and innovation. Instead of radical transformation the triad will involve top-down iterative change where environmental and economic benefits are produced simultaneously – a so-called win-win transition. 3D printing in this model will emerge as ubiquitous due to its capacity to underpin a renaissance for manufacturing in the Global North's production complexes, where there is a return or 'reshoring' of manufacturing from current regional clusters in the Global South. 3D printing reinvigorates post-industrial regions' economies and secures employment growth in both the material and knowledge economies, which converge around hybrid nodes. As a method for increasing flexibility and leanness in distribution 3D printing will become just another part of existing supply-chains, making them more 'green' and ultimately more responsive to consumers' needs and wants. Finally, both online and offline retailers will blossom through suppliers of products meeting consumer demand for variety and choice through mass customization and print-on-demand business models.

Here neoliberal governance in the Global North will thrive on economic growth from escalating employment, consumer satisfaction, and the meeting of environmental targets. Governance trends in the Global South will compensate for the loss of mass manufacturing clusters with the rising tide of incomes, education, cosmopolitanism and standards of living anticipated from developing knowledge and services economies. Notwithstanding the challenges in decoupling production from distribution and other instances of energy-use, there are many oversights in this understanding of transition stemming from substituting resources, chiefly fossil fuels, with sustainable alternatives while maintaining economic growth across the world. Reform in this model depends on 3D printing being a 'technological fix' – an evolutionary progression of technology, which current assessments of its affordances reveal is problematic at best.

Second there is the 'classic steering' notion of governance wherein driving a 'revolutionary' position to socio-technical change. Non-governmental organizations, social movements and radical scientists often embody the position. Involving a radical critique of neoliberal governance, 3D printing in this instance is disruptive and epochal, resulting in the collapse of capitalism, materialism, and over-consumption. In place 3D printing underpins values such as frugality, sufficiency, and localism. Informal economies and even currencies emerge from the periphery as alternatives to consumer capitalism and 'big brand' object producers. Maker communities and spaces motivate more thoughtful and engaged consumers whose scrutiny of object production and distribution standards undermines the current triad. 3D printing allows many of these consumer-citizens to go 'off-grid' and make their own objects themselves from recycled or renewable materials and shared 'open source' digital designs with government stewardship and subsidization.

Here progressive governance in the Global North will destabilize the production–distribution–consumption triad in concert with political movements already in place around the world, for instance the so-called 'pink tide' in Central and South America (Cuba, Venezuela, Brazil, Uruguay, Bolivia, Ecuador and Nicaragua) or Cuba. Opposition to this model will occur on multiple fronts, from consumer groups resenting or resisting a possible curtailment of access to low-cost, disposable objects to the super-rich who profit from both offshoring and their business interests. Corporations unable to reconcile their shareholders' demands for wealth creation will lobby and fund government parties' campaigns against this revolutionary position, with consequent support from consumer-citizens.

As Geels and colleagues note, there are many roadblocks to either of these two futures of reform or revolution. In response they propose a third way: 'network governance', currently only visible at the city level by authorities who enact transitions with an eye on both overarching systems and individuals' social practices within them, for instance the London congestion charge scheme.[31] It is not a question of changing what individuals do or do not do, but of changing whole systems made up of various elements: economic ones, technological ones and social practices. Systems are crucial here and not just individual behaviour.[32] Such a reconfiguration position focuses on transitions in socio-technical systems and daily life practices, and accommodates a new conceptual framework as an alternative to approaches that push for reform or revolution.

## Conclusion

Thus we might conclude that 3D printing requires more rather than less engagement between people and machines to reach a level of ubiquity that would have consequences for the current system of extraction, production, distribution, consumption and destruction. Rather than becoming one-dimensional, in Marcuse's understanding, people could become 'three-dimensional' as actors engaging in producing the objects they use and collaborating with others around them. Massification on the one hand provides people with the means to sustain the historically unprecedented surge in the human population and consequent demands on its resources. On the other hand, massification has associations of disposability, superficiality, homogeneity, blandness and environmental destruction. In the next chapter we consider the empirical method of social futures in order to provide a scaffold for the scenarios in the final chapter.

## Notes

1 M. Raby, '3D Printing Market to Hit $3 Billion by 2018'. New York, 2012. Accessed 3 August 2012. http://www.slashgear.com/3d-printing-market-to-hit-3-billion-by-2018-23239870/

2 R.A. D'Aveni, '3-D Printing Will Change the World', *Harvard Business Review*. Harvard: Harvard Business Publishing, 2013. Accessed 12 January 2015, http://www.hbr.org/2013/03/3-d-printing-will-change-the-world/

3 Z.B. Ilan, 'From Economy of Commodities to Economy of Ideas: Hardware as Social Medium'. *Design Management Review* 22, no. 3 (2011): 44–53, doi:10.1111/j.1948-7169.2011.00139.x. 48

4 J. Woudhuysen, 'Drilling into 3D Printing: Gimmick, Revolution or Spooks' Nightmare?'." De Montfort University, 2013. Accessed 19 March 2013. https://www.dora.dmu.ac.uk/xmlui/handle/2086/8696

5 M. Holweg, 'The Limits of 3D Printing', *Harvard Business Review*. Boston, MA: Harvard Business Publishing, 2015. Accessed 4 August 2015. https://hbr.org/2015/06/the-limits-of-3d-printing

6 V. Tang, 'Australia's First Commercial 3D Printing Factory Opens in Melbourne', 2015 Objective3D. Accessed 4 August 2015. http://objective3d.com.au/australias-first-commercial-3d-printing-factory-opens-in-melbourne/

7 A. Leonard, *The Story of Stuff*. New York: Simon and Schuster, 2010.

8 S. Luckman, *Craft and the Creative Economy*. Basingstoke: Palgrave Macmillan, 2015.

9 M. Holweg, 'The Limits of 3D Printing', *Harvard Business Review*. Boston, MA: Harvard Business Publishing, 2015. No pagination. Accessed 4 August 2015, https://hbr.org/2015/06/the-limits-of-3d-printing

10 T.A. Hutton, 'Reconstructed Production Landscapes in the Postmodern City: Applied Design and Creative Services in the Metropolitan Core', *Urban Geography* 21, no. 4 (2000): 285–317, doi:10.2747/0272-3638.21.4.285

11 D.E. Nye, *Technology Matters : Questions to Live With*. Cambridge, MA: MIT Press, 2006.

12 J. Urry, *Climate Change and Society*. Cambridge: Polity Press, 2011. p. 54.

13 G. Simmel, *Sociology: Inquiries into the Construction of Social Forms*. BRILL, 2009.

14 T. Tokuda, *Cars in the '90s as a Humanware*. Warrendale, PA: International SAE, 1988. SAE Technical Paper 885049 10.4271/885049.

15 J. Urry, *Societies Beyond Oil: Oil Dregs and Social Futures*. London: Zed, 2013.

16 T. Birtchnell and J. Urry, 'The Mobilities and Post-Mobilities of Cargo'. *Consumption Markets & Culture* 18, no. 1 (2015): 25–38, doi:10.1080/10253866.2014.899214

17 K. Dennis and J. Urry, *After the Car*. Cambridge: Polity, 2009.
18 E. Griffin, *A Short History of the British Industrial Revolution*. New York: Palgrave Macmillan, 2010.
19 A. Comte, *The Positive Philosophy*. New York: AMS Press Inc., 1974.
20 E. Durkheim, *The Division of Labor in Society*. London: Free Press, 1997.
21 K. Marx, *Capital: A Critical Analysis of Capitalist Production*, Vol. 1. London: Sonnenschein, 1887.
22 N. Hopkinson, R.J.M. Hague and P.M. Dickens, 'Introduction to Rapid Manufacturing'. In *Rapid Manufacturing: An Industrial Revolution for the Digital Age*, edited by Hopkinson and Hague, 1–4. Chichester: John Wiley & Sons, 2006.
23 P. Andreev, I. Salomon and N. Pliskin, 'Review: State of Teleactivities'. *Information/ Communication Technologies and Travel Behaviour; Agents in Traffic and Transportation* 18, no. 1 (2010): 3–20, doi:10.1016/j.trc.2009.04.017; P.L. Mokhtarian and I. Salomon, 'Emerging Travel Patterns: Do Telecommunications Make a Difference?'. In *Perpetual Motion: Travel Behaviour Research Opportunities and Application Challenges*, edited by Mahmassani, 143–82. Oxford: Elsevier Science Ltd., 2002.
24 A. Toffler, *The Third Wave*. London: Collins, 1980. p. 198.
25 E. Shove, M. Pantzar and M. Watson, *The Dynamics of Social Practice: Everyday Life and How It Changes*. London: Sage Publications Limited, 2012.
26 U. Beck, 'Rooted Cosmopolitanism: Emerging from a Rivalry of Distinctions'. In *Global America? The Cultural Consequences of Globalization*, edited by Beck, Sznaider and Winter, 15–30. Liverpool: Liverpool University Press, 2003.
27 S. Luckman, *Craft and the Creative Economy*. Basingstoke: Palgrave Macmillan, 2015.
28 G. Graham, R. Mehmood and E. Coles, 'Exploring Future Cityscapes through Urban Logistics Prototyping: A Technical Viewpoint', *Supply Chain Management: an International Journal*, 20, no. 3 (2015): 341–52, doi:10.1108/SCM-05-2014-0169. 347
29 M. Potstada and J. Zybura, 'The Role of Context in Science Fiction Prototyping: The Digital Industrial Revolution'. *Technological Forecasting and Social Change* 84 (2014): 101–14, doi:10.1016/j.techfore.2013.08.026. 109
30 F.W. Geels, A. McMeekin, J. Mylan and D. Southerton, 'A Critical Appraisal of Sustainable Consumption and Production Research: The Reformist, Revolutionary and Reconfiguration Positions'. *Global Environmental Change* 34 (2015): 1–12, doi:10.1016/ j.gloenvcha.2015.04.013
31 E. Shove and G. Walker, 'Governing Transitions in the Sustainability of Everyday Life'. *Research Policy* 39, no. 4 (2010): 471–76, doi:10.1016/j.respol.2010.01.019
32 J. Urry, 'Sociology Facing Climate Change'. *Sociological Research Online* 15, no. 3 (2010): 1.

# 6

# FUTURITY

## Introduction

The overarching framework for this book's scenarios is a world defined not by nation-states but by mobilities.[1] Critical mobilities research provides an alternative to a regional optic through radically revising understandings of the production of space; the politics of transport; the fluid hypermobility of elites; the (im)mobilities of the still, stranded and stateless; and the relations between bodies, movement and space more generally.[2] Mobilities research focuses on what moves – people, vehicles, information, and objects – as well as systems that enable movement – roads, servers, cables, pipes, logistics and queue management technologies.[3] Taking mobilities into account, there are two interlaced drivers of social transformation which are significant for forecasting the 3D society. These drivers are, notionally, fossil fuels, climate change and governance.

The first uncertainty of the 3D society in this book is energy. Fossil fuels – petroleum, coal and natural gas – are unequivocally the energy source par excellence for the majority of the world's transportation technologies.[4] As explained earlier in this book mass manufacturing is centralized in clusters of factories in a handful of regions. A by-product of this centralization is future risk arising from high-energy costs due to the continuing need for growth in bulk forms of production and transoceanic shipping. 3D printing minimizes the risk of disruption as objects made close to or even by consumers reduce the need for transportation energy and improve efficiencies in material conservation and disposal. For some products the affordance of 3D scanning an object and then making multiple copies would produce large cost savings and reduce transport-related fuel costs. The consequence of mitigating risk of disruption in energy could mean that at some point the low-cost manufacturing clusters, in the Global South, would no longer possess a comparative advantage in production and distribution over thousands of miles to the Global North. As 3D printing is a non-transport technology outside of 'transportation' the uncertainty over the future of fossil fuels could be managed by eliminating the mass containerization of many manufactured objects. A 3D society could perhaps even eliminate the whole discipline of 'logistics'. Digital objects can travel almost for free through optical cables and wireless masts, although fossil fuels are the basis of many of the feedstocks in this kind of additive manufacturing.

The second uncertainty of the 3D society, related to the first, is the climate.[5] For all scenarios we assumed that climate change would be an important factor. 3D printing innovations offer possible futures of rapidly demobilizing global manufacturing, distribution and production. Already 3D printing has featured in a Delphi panel on the future of air cargo. This research envisaged a 'wildcard scenario' called the 'Fabbing Society' where personal home fabricators and decentralized additive manufacturing facilities combine to wreak havoc on the existing air-freight industry.[6]

The third uncertainty of the 3D society is governance: who will be in charge in the future and how this will shape the 3D printing ecosystem. Today many would say neoliberalism holds sway, and this is certainly the case in the US, UK, Australia and Canada. Yet elsewhere in the world are alternatives to a neoliberal worldview, notably in Germany and the Scandinavian states, where social liberalism prevails at the same time as economic growth and high standards of living. The flavour of government will have an impact upon the other two uncertainties, with energy and climate policies polarizing depending on who is in power.

Fourth there are global social inequalities. As well as upturning the global production network there is also scope for some levelling of income inequalities and tackling poverty through 3D printing. An ambitious project by entrepreneur Kartik Gada foresees personal manufacturing substituting for aid and charity in reducing poverty. Gada created the 'K Prize' for innovators in 3D printing to encourage low cost manufacturing and reduce the 'fixed costs and volume necessities associated with manufacturing' so that 'the scale of Chinese mass manufacturing is no longer a requirement to be cost competitive'.[7] The prize money will go to the innovator able to produce a fully self-replicating printer that can make 90 per cent of its own parts.

## Engagement with 3D printing

What people and organizations do with a given innovation is decisive in its ubiquity and mass adoption. A number of engagements are happening now. First, rapid tooling involves the production of limited numbers of moulds for the traditional process of factory injection moulding, shortening the production time for small numbers of high value parts, such as for nuclear submarines. Similar to rapid prototyping, rapid tooling is invisible to the consumer beyond providing them with more options for custom products.

Second, beyond models and moulds, direct manufacturing is another way by which 3D printers are emerging in new business models. The aerospace and automotive sectors already use 3D printers in this way. In the former, 3D printing enables so-called mass customization: unique steering wheels made to fit their owners.[8] In the latter, 3D printing offers geometric complexity that is impossible to achieve any other way: hollow bird-bone-like structures for aeroplane wings.[9]

Two candidates are obvious for direct manufacturing: DMLS uses a laser to weld beds of powdered metal into structures; EBM offers a more powerful (higher energy density) form of computerized welding, although this technique is generally understood to be more expensive and time-consuming. Obviously direct manufacturing also modifies distribution and transoceanic freight systems, as products are simply made closer to where they will be used. Taking these current uses of 3D printing on board, analysts recognize that in future its use will be driven by the 'co-evolution of technology expectations and technology development activities', namely that there will be an interplay of consumer demand and supplier innovation.[10]

Beyond integration into industrial systems another current use of 3D printing is in creative practice: in jewellery, fashion and for artworks. On 15 April 2015 Routledge published the edited book *Cargomobilities: Moving Materials in a Global Age*.[11] Partly inspired by this book collection, an installation with the same name was exhibited in the Art Center College of Design's Hutto-Patterson Exhibition atrium in early June of 2015 in the US by artist Joelle Dietrick (see Figure 6.1). The first in a series, the installation was a prototype for a second installation at the Project Atrium, Museum of Contemporary Art (MOCA), in Jacksonville in late June, also titled Cargomobilities. Learning of the installations on the social media

**FIGURE 6.1** Complex collage of cranes and cargo. Joelle Dietrick Cargomobilies (Los Angeles) at Art Center College of Design, Pasedena, California, June 1–September 1, 2015.

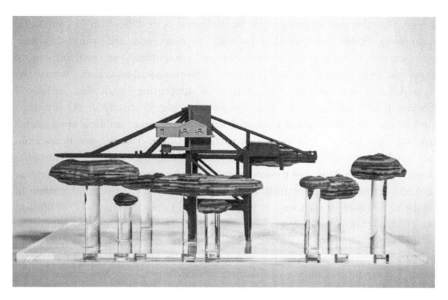

**FIGURE 6.2**   3D printed model of Joelle Dietrick's Cargomobilities installation by
Thomas Birtchnell, Joelle Dietrick and Robert Gorkin.

platform Twitter via the hashtag #cargomobilities, we were inspired to create a
post-exhibition 're-make' using a 3D printer as a creative collaboration in the conceptual space opened up between the book and the artwork, drawing on digital
artefacts of the original shared online.

Dietrick's Cargomobilities installation was an impermanent work of art, removed
and whitewashed after the exhibition, its ghostly digital residues remaining only on
the Internet and in viewers' minds. Using a 3D printer the remake from digital
artefacts offered a visual commentary on the book's major themes as well as allowing viewers to experience the installation in three dimensions, albeit at a different
scale (see Figure 6.2).

The collaboration shows the benefit of 3D printing for creative practice, and in
this case the technology was used for a post-production rather than pre-production
purpose as the model came after the 'final' product. It is these kinds of applications
of the technology that suggest new and different kinds of relationships between
people and production. In light of both the above creative and industrial – that is,
rapid prototyping, rapid tooling and direct manufacturing – engagements with 3D
printing, a similar crucial uncertainty will be the openness of the technology either
as something that works behind the scenes or as something people actively integrate
into their lifeworlds.

## Openness of 3D printing

The second uncertainty in 3D printing's emergence as a socially consequential
innovation is the degree of 'openness', an inference that can be drawn through

inquiry into the level of profit made by corporations or individuals in a particular technology. Traditionally, social, economic, environmental, and spatial data have been generated and held by governments and companies, who own and control copyright, and restrict its use and purposes – however social movements who lobby to make data free cite the role of taxes in supporting much data creation.[12] Commentators emphasize the role openness has played already in 3D printing's current success from consumer-focused start-up companies utilizing open source hardware and software frameworks. In part, 3D printers have become mainstream due to the introduction of these open-source technologies onto the consumer market; the recent demise of patents for metal 3D printing suggests further innovation, marketization and systematization of the technology for a broader range of products.[13] A spate of start-up companies now offer consumer-level FDM units to the public drawing on freely available designs, notably the 'Reprap'.[14] An example of this is the company Makerbot whose flagship consumer-level for-profit 3D printers, including the Replicator, profited from the open source FDM hardware of the Reprap project and the open source software repository Thingiverse offering free downloadable content alongside for-profit design files.[15]

A leap forward for open source innovators has been to use generic stocks of raw materials, readily available to consumers to purchase or accumulate. Of issue here is the diameter of the filament 'wire' in relation to the extrusion head. By using generic feedstock, inventories are greatly reduced, as stockists only need to order a single kind of product that can be used in a wide range of different printers. Hence in the use of generic feedstock there is potential for efficiency gains in transport and storage.

Similar to the computer revolution there are entrepreneurs, mergers and acquisitions – Stratasys's takeover of Makerbot is a notable milestone. Also similar to the computer revolution, there is a sizable cohort of hobbyists, tinkerers and radical innovators whose activities are not motivated by profit or earning a livelihood. One recent example is the Reprap 'open source' 3D printer. An underlying theme is that the ability to print objects could extend to the printers themselves. The capacity to do away with the market system altogether in favour of 'wealth without money' has been dubbed a 'Darwinian Marxist' revolution.[16] More pragmatically, according to inventor Bath University's Adrian Bowyer, the Reprap is 'designed to copy itself because that's the most efficient way of getting a large number of them out there'.[17] If 3D printing does enter the production cycle for end-user products, this will introduce a whole new dimension to debates about sharing, open source and piracy. For instance, even in an apparently 'open' repository such as Thingiverse only a proportion of users license their content in ways that take full advantage of open licensing norms.[18] The regulation of 3D printing will involve overcoming many challenges, which currently would be non-sequiturs: is a CAD file a 'product' for instance?[19] These are more social than technical problems.

Because 3D printers use digital design files blueprints can be traded in the same way as digital music and text. Websites including Thingiverse, Cubify, The Pirate Bay – which offers a 'physibles' category – and Instructables are already catering for this new demand for filesharing of designs. The scope for reverse engineering is already on the horizon. Multiple colours are available thanks to the mixing of different powder colours in high-end models. More expensive printers also use binding agents and powders, lasers and electron beams, and exotic materials such as resin, nylon, plastic, glass, carbon, titanium, sand or stainless steel. The latest machines can also mix together a number of materials in the same print and produce durable parts for cars, bikes and planes.

On the one hand, two possible visions are imagined in 3D printing discourses around uncertainties in openness – a utopian one where 3D printing combines well with existing encryption systems to develop radical economies outside of the current regulatory capabilities, and a dystopian one where the openness of 3D printing combines with a lack of scarcity for dangerous items, leading to a proliferation of 'fascistic organisations, terrorists, and sexual perversion'.[20] Such issues are wrapped up in concerns around the commons and public access to intellectual property, innovations and data more broadly. There are social ramifications as consumer production in a distributed fashion is enabled.[21]

On the other hand 3D printers could be seamlessly integrated into existing production or supply chains with little to no openness for consumer-citizens.[22] Relationships between manufacturers and various subcontractors for product parts are one area of change where openness is not required, as are inventories and the use of 3D printing in storage and handling. Fewer tasks and fewer steps will serve small markets of bespoke products suitable for customization, and here 3D printing could operate in a closed environment.

There are considerable legal complexities around 3D printing from files downloaded from the Internet as with music and text. US patent attorney Daniel Harris Brean notes that under existing law distributors of digital representations of products, such as CAD files, are not 'making', 'selling', or 'using' the products or any 'component' thereof. Indeed, the legal implications of 3D printing are not clear-cut and could entail black (or at least grey) swans for policymakers.[23]

However, most of the excitement and anticipation focuses on business models transforming consumers in their homes into producers, or 'prosumers', involving the profit from co-creation, crowd sourcing and other user innovations.[24] A determining factor will be the role of interactive and easy-to-use platforms that realize designs and ideas into objects without the loss or violation of intellectual property rights.[25] In contexts of resistance to openness, market structure models will require forethought, precision, and in some cases the surveillance and control of citizen-consumer activities.[26]

Parallels are already being drawn with the music industry and the proliferation of piracy online in the wake of the digitization of object data.[27] Furthermore, even online 3D printing file repositories, such as Thingiverse, do not 'practise what they preach', with many users making their files private and not publically accessible as the culture of sharing promises.[28] Undoubtedly, patent enforcement will prove to be an extant issue for businesses to make a profit in future.[29] In light of this uncertainty there is likely to be a shift to a focus on information management as the cost of 3D printing diminishes for consumers with greater digitization and sharing of openly available data. One other factor is how much consumers will want to engage with production processes (open or not), with one study finding ambivalence amongst entry-level users due to the quality, cost, time, and patent restraints.[30]

## Futures workshop

To examine the four uncertainties outlined in this chapter this book draws on a scenario-building exercise held in 2012. Why use scenarios? The use of scenarios in futures studies arose due to efforts to offset the uncertainty that is inherent in single-track, long-range forecasting.[31] By drawing on broad sources of data 'out in the world', ranging from the media and public opinion to scientific consensus and modelling, scenarios attempt to avoid the bias that religious divination suffers.[32] Scenarios do not claim to be able to interpret the future and in this way lend balance to forecasting. The development of multiple scenarios in a scientific, rational manner arose in the mid-twentieth century to cater for the demand for objectivity and a 'science' of forecasting.[33] Scenarios are useful for objectivity but not so much for subjectivity. In scenarios there are limited opportunities for prosaic forecasts pertaining to individuals and their own lives. No-one ever made their fortune from making scenarios. Instead, scenarios find purpose and application in developing policies, laws, proposals and responses to broad challenges and opportunities.

Multiple scenarios are also useful to develop fuller visions that represent 'preferred futures' for influencing decision-makers, investors or communities of citizens.[34] Scenarios are not figments of future realities per se – sent back in time like snapshots – they are ways of foreseeing trends towards futures that throw light on the present. They are also liable to be unreliable and even failures. Scenarios are neither necessarily going to come to be familiar nor will they allow for unintended consequences, radical innovations, new ideas or large-scale political transitions (as much Cold War futurology demonstrates).[35]

It is no coincidence that the idea of progress, particularly in terms of industry, has historically been a common theme in scenario-based forecasting. Critics contest that much of this focus is techno-optimistic. Indeed, the idea of technological progress has encouraged an overly simple dialectic over the lifetime of recent futures studies.[36] In the mid-twentieth century progress, in terms of technological development, foresaw 'utopic' images of machine-made bounty, prodigious leisure, and boundless human enterprise. In the mid-twentieth century progress was a product of a 'modern' evolutionary understanding of technology. In the twenty-first century the 'post-modern' idea of progress shifts to images of sustainable human and economic growth and collaboration, rather than evolutionary competition, with the environment and its diverse biomes. In this utopic revision of the modernist idea of technological development, revolution rather than evolution takes precedence.

Partly, the shift from modernist to post-modernist forecasting is due to the many scenario forecasts of dystopic visions that the late twentieth century saw developing. Influencing these scenarios were Malthusian scientific appraisals of the future that still hold water today. Predicted impending catastrophes from overpopulation, climate change, resource depletion, and environmental extinction and collapse, are part and parcel of scenarios. A solution to overly simple techno-optimism is forecasting that combines predictions of technological development with what are historically 'neglected' aspects in scenarios. Attention to the extant features of 'socio-technical' transitions problematizes conceptualizations of technological development and its impact on society in terms of utopias and dystopias. Accordingly, an overt focus on the impact of new technologies without the social in mind, in terms of either the direct consequences or the indirect ones, is a source of imprecision.[37]

The style of scenarios used in this book is exploratory in order to capture the variety of possible societies involving 3D printing. We drew on a standard methodology in future studies in the development phase: two dimensions of change within which four scenarios were constructed around a two-by-two matrix. Quadrants were independent, equally likely futures, each with a unique theme and logic that describe the content.[38] In structuring this approach, the power of the central variables was exaggerated so there was space for creative design and 'mind stretching'.[39] The approach involved scripting narrative structures within the limits of the matrix in order to present futures as demonstrably self-contained and thematic, similar to refining a narrative in a story.

Alongside the formal trends for each future were vignettes we composed in order to inspire the workshop participants. The important role of creativity in futures work is not only found on the fringes of debate. Fictional stories have featured with good effect in the UK Government's Foresight reports. One of the authors was involved in a novel approach to assessing the possible futures of transport and mobility in one of these exercises. In the report 'Intelligent Infrastructure Futures – Towards 2055', four scenarios of different worlds were explored and in each a fictional vignette was integrated into the workshop. This was called 'An Urgent Delivery' whereby the character 'Mike' attempts to deliver a package in the radically different worlds imagined within each scenario: 'He switched the system to robot mode – accepting that he would make 20% less money this afternoon – locked the control room and walked across the park to the cyberpub which had its own micro-brewery and a secluded garden where use of wifi equipment was forbidden'.[40] Creative approaches to scenarios are now recognized as both a valuable asset for thinking about the future in general terms, and a possible method for specific inquiries into how people engage with technologies and the degree to which business shapes people. Fictional devices enhanced this research into 3D printing and its potentially large-scale consequences.

Developers of scenarios acknowledge that there is a creative element to the craft of forecasting: the process is as much an art as a science. Similar to a fictional world, scenarios are unlikely to ever occur in reality: 'a scenario with 20 variables, assuming that they are independent and each was 90% certain, the likelihood of the scenario based around it occurring would, on a strictly mathematical basis, be about 12%'.[41] To mitigate too much creativity, experts emphasize that the fictional elements in forecasting are made transparent for credibility and usefulness.[42]

This use of creative fiction in policy and planning contexts, in order to imagine the social aspects of future worlds, has subsequently appeared in other related academic scenarios work we have been involved in relating to future transport systems, climate change and oil depletion. In this book we describe similar experiments relating to the current state of play of 3D printing and its possible future impacts upon the movements of objects and people. We link the development of 3D printing to fictional futures and highlight the possible ways these visions might interestingly inform forecasting and scenario building. We also highlight the invaluable role creative thinking played in examining scenarios and collaborating with experts within engineering, design, consultancy, and policymaking.

Beyond creativity there are three core elements in the scenario typology, regardless of whether it is exploratory or not, that serve as the mechanics of the exercise.[43] In the case of this book the characteristics of the project goal are descriptive, exploring possible futures in order to support policy counsel and intervention. The study was issue-based in its focus on individual engagement and corporatization. The time-scale was to the year 2050. The design involves the exploration of situations or developments that are regarded as possible to happen from a variety of perspectives.

The 'Ethnographic Futures Framework' (Verge), developed by Kaipo Lum and Michele Bowman, raised questions and vectors of analysis in the first stages of the workshop and encouraged thinking outside of established norms.[44] The ethnographic futures framework focuses on social transformation through human agency and utilizes five categories to explore change: 'define' (how will people define themselves, what concepts, ideas and paradigms will emerge to help them make sense of the world?); 'relate' (how will people relate to each other and the world around them?); 'connect' (what media and technologies are used to connect people and places?); 'create' (how will people create new goods, services and knowledge?); and 'consume' (how will people use and dispose of resources?).[45] These categories were modified for the project.

Four different groups in the workshop explored how in each world people's needs in the built environment might be different and how policymakers would react according to a framework characterized by seven descriptors. *Create*: What is produced in this scenario? How is it produced? Why is it produced? *Consume*: What do people consume? Where? Why? How do they think about resources? *Destroy*: How do people dispose of materials when people have finished with them? *Connect*: How do we connect with people at a distance? What communications technologies and networks are important? What transportation systems do people depend on? *Relate*: How do we live together? What are our (physical) communities like? What are the most important social relationships? What sorts of organizations express our social values? *Define*: What concepts, ideas, and paradigms inform the way people understand the world in this scenario? *Transport*: What are the transport and mobility implications?

A marked shift to demand for customization guided the participants' response to the first category 'define'. In a world of infinite choice and variety – one that 3D printing promises – the idea of the prosumer (a portmanteau of producer and consumer) or produser (a portmanteau of producer and user) will enter the mainstream. If no two objects need be alike once 3D printing reaches ubiquity, how will people define themselves in terms of identity? Will nostalgia emerge for mass manufactured objects that feeds into resistance movements and backlashes to variety? A greater resolution in the discernment of difference will manifest in the perception of objects: a forest might appear to be made up of trees which are all alike, but on closer inspection each tree in fact has its own character.

With 'relate' the participants foresaw much collaboration and at the same time competition in gaining access to inimitable designs of quality. Printers might be accessible to communities or just to individuals.

After a review of outcomes, the groups used a second tool – the Technology Axis Model 2 – to deepen their understanding of potential points of disruption within each scenario. The Technology Axis Model helps analyse the impact and uptake of emerging technologies – in this case 3D printing. This enables a socio-technical approach to be taken, understanding both the enabling technologies and the infrastructure needed for the technology to evolve rapidly, as well as the social values and practices that frame its development.

The Technology Axis Model is designed to consider the impact of each of the four elements: applications, societal norms, technology and systems. It can be read in either a clockwise or counter-clockwise direction: there is no starting point. It is not deterministic: there is feedback and interaction between each element. For the purposes of the workshop, the model was applied to each of the four scenarios.

The scenarios in this book were refined and tested through a workshop held in London with 24 experts. Experts were drawn from the academic community, business and the public sector, and from the areas of design, consultancy, engineering and policymaking. Together we identified the key drivers and trends for 3D printing in order to produce possible futures. The axes of uncertainty framing the project are whether or not individuals will engage with the socio-technological ecosystem, and whether or not 3D printing will be communal or corporatized.

## The scenarios

In this book a four-scenario framework is drawn from two axes of uncertainty. The polarities give rise to independent scenarios, each with its own 'flavour'. The use of axes to produce social futures derives from the work of sociologist Daniel Bell who pioneered the approach in *The Coming of Post-Industrial Society: A Venture in Social Forecasting*.[46] In this landmark book Bell aimed to analyse the character of a society through the heuristic of axial principles that sought to avoid the pitfalls of prediction by specifying the limits within which policy decisions can be effective in terms of changes in social frameworks.[47] The approach differed from other social future methods from the time by drawing on multiple, richly informed perspectives.

Yet the study is also a product of his time, one in which social scientists would specialize in distinct regions and compare them. Before studies of globalization took hold nations were understood as fixed, closed and containing very different cultures and peoples with neither much contact nor commonality. Informed by this optic Bell's two axes were the polar extremes of pre-industrial to industrial and collective to capitalist. Two key trends presented themselves to Bell as orthogonal: the extent of technological innovation and the privatization of property. Usefully for him there were four precisely demarcated regions that he understood as following these polarized trajectories. The four regions were either congruent or antagonistic depending on the axis of comparison: Soviet Russia, America, China, and Indonesia.

Usefully for this book's methods, Bell's venture – in examining the futures of post-industrial societies – is in the same ball-park as 3D printing, however, we did not establish the axes of this study around uncertainties in the development of regions. For Bell the US is the archetypal capitalist form of society on a fundamentally opposite trajectory to that of communist Russia. Indeed not all were satisfied at the time with Bell's regional approach.[48] With hindsight it is challenging to relate the research to current events, despite its future-facing stance. As we now know with the advantage of time, Bell's scenarios make little sense with the collapse of the Soviet Union, the People's Republic of China's dramatic economic growth, and the emergence of both as capitalist super-powers – each according to their own unique flavour – to rival the US in technological innovation and privatization of property.

There is little to gain from a belaboured criticism of Bell's choices of regions for quadrants. Instead we build on his venture by benchmarking this book's research against recent uses of the so-called 2x2 method in the social sciences and government policymaking. Instead of focusing explicitly on regions, we draw in this book on the mobilities paradigm to move beyond the nation-state demarcations that Bell foresaw as propitious for his research. To mitigate some of the shortfalls of the process, we drew on accounts from a specific research exercise with experts: a 'futures' workshop held to guide this book's foresight. The focus was multi-scalar, with drivers evident not only at the nation-state level, but also across economic 'classes', sub-cultures, alternative social movements, and so on.

The first axis of uncertainty underlying the four quadrants is akin to the concerns of Bell's post-industrial society thesis: will the future be more corporate or open? To put it another way, will a 3D society involve the further consolidation of global resources by a handful of elites through neoliberal and free market ideologies in government and industry? Or will the 3D society pivot off of a turn to democratization, transparency, and equal access to – and distribution of – technology, welfare and wealth. Distinct phenomena within the 3D printing ecosystem offer an intimation of these polar opposites.

Already there are signs of corporate monopolies emerging through various mergers and takeovers. The two major players in the world in terms of investors and patents are the corporations Stratasys and 3D Systems.[49] The companies were founded respectively by heroic innovators – similar to Steve Jobs and Bill Gates in the personal computer sector – Charles (Chuck) Hull and S. Scott Crump, with each championing different techniques and innovations including SLA and FDM. In mid-2012 both corporations announced impressive mergers, signalling efforts to dominate the 3D printing market. Stratasys merged with Israeli company Objet, boasting almost 500 patents and their proprietary PolyJet technology, at the time the only multi-material printer, useful for both personal and industrial applications. 3D Systems made two acquisitions: My Robot Nation and Paramount Industries. The first targeted personal 3D printing, chiefly the consumer market. The second company was a larger player in aerospace and medical. The two acquisitions also positioned 3D Systems as a leader in the personal and industrial markets.[50]

At the other extreme, the Reprap, discussed earlier, is the flagship of a burgeoning and global social movement in 'peer production'.[51] Through the co-creation of a 'digital commons' these movements draw on distributed expertise and, more often than not, volunteers' time and energy. Users are also co-creators in providing testing and feedback. Poignantly, the spread of the Reprap framework is spawning a spate of start-up companies utilizing its open source design in order to market pre-assembled, affordable, consumer-level 3D printers. The most successful to date, Makerbot, was – perhaps ominously – purchased by Stratasys for US$604 million in 2013, demonstrating that open source technologies can also be profitable.[52] Whether the 3D society is corporate or communal by nature will be dependent on many variables.

The second axis of uncertainty in the study is the level of individual engagement with the technology. The 3D society has many parallels with the personal computer revolution, however it is not a given that ubiquity will result in people using the devices in their homes. While it is certainly the case that many SLA and FDM printers are now available for consumers to use in their homes, this is not so for the more industrial species of printers. Early findings from a research project being conducted by the Additive Manufacturing and 3D Printing Research Group at the University of Nottingham, and Saïd Business School at the University of Oxford, point to both significant scale and learning effects inherent in industrial selective laser sintering and electron beam melting.[53] It might be that individuals are never in a position to engage with the affordances of the more complex 3D printing eco-system. Bureaus, fablabs and other institutional collectives will be the nearest they get to these kinds of productions. So to summarize, the uncertainty derives from a number of aspects.

The technical features of the 3D printers themselves are relevant here. There is much uncertainty about whether individuals will in future access industrial-scale 3D printers individually or communally – laser sintering or electron-beam melting systems – unless the technology develops in the same kind of trajectory as personal computers (in earlier phases of development these were restricted to 'expert' users). As Chapter 1 summarizes, many of the features of the technology will be incompatible with a home environment. In fact, this is a point of dispute in much of the commentary. For instance, in a recent video interview engineer Richard Hague, from the University of Nottingham, noted 'one of the misconceptions about additive manufacturing is that every home will have one of these 3D printers; that doesn't reflect reality'.[54] Hague goes on to suggest an alternative transition pathway for 3D printing's ubiquity, the 'printing photographs at home model': users create the digital data, but utilize external organizations to print the objects. Hague's point is compelling. Many of the 3D printing innovations – beyond the heating and extrusion of plastic – currently in the research and development phase offer affordances to consumers far beyond the current mass manufacturing system, yet it is unclear how compatible these would be in terms of safety, size and cost in the home environment. It is likely, however, that the next generation of 3D printing technology would be compatible with the sorts of digital files that those on the market today use. Complicating the axis somewhat are different notions of individual engagement.

An uncertainty in the ubiquity of 3D printers is the learning curve and cost of software systems for creating or manipulating digital designs. There are many templates, repositories and WYSIWYG (what you see is what you get) applications for those lacking technical skills. Moreover, many consumers would simply not want to customize or create their own objects beyond the choices they currently get in the current system: colour, size, brand, and so on. Designer Lisa Harouni offers a useful survey of the possible types of engagements individuals will have in the future. Already, users are re-engineering designs made by professionals for their own purposes. Harouni is candid, however, that software is unlikely to substitute directly for professional designers.[55]

Overall individual engagement with the 3D printing ecosystem is liable to be either low or high depending upon many variables. Evidence for this includes a survey submitted online to 170 users from 30 different 'makerspaces' (cooperative community manufacturing networks) in more than ten countries. The study demonstrates that even in these peripheral spaces of high individual engagement – made up of inexpensive, mostly 'extrusion' units – the use is variable. On the one hand there are 'expert' users who prototype for testing or to make designs for more complex printers, and on the other hand 'beginners' printing objects they can use themselves.[56]

The first quadrant, 'Print-It-Yourself', combines an open, unrestricted ecosystem with ubiquitous individual engagement in 3D printing. In a society where openness is neither maverick nor an aberration of 3D printing alone, neoliberalism is shunned in favour of social liberalism. People are 'doing' 3D printing themselves and sharing what they do with aplomb; there is little regard for intellectual property, copyright or licensing. The first quadrant involves 3D printers reaching social saturation through self-replication, so much so that they are familiar and a common sight in people's homes or in the nearby vicinity. In fact people's homes are 3D printed, allowing a great deal of architectural freedom. The transition is in the same fashion as wristwatches in the 1920s, refrigerators in the 1930s, automobiles in the 1950s, personal computers in the 1980s, mobile phones in the 2000s, and smart phones in the 2010s. In short, 3D printers become everyday, household objects, either left in place to operate as needed in kitchens and studies or carried about on the person in a miniaturized form. The 3D printing ecosystem reaches ubiquity in this quadrant through a wider drive towards global sustainability and the massification of off-grid energy.

The second quadrant, 'I Print Therefore I Am', involves mass individual integration with a corporatized 3D printing ecosystem mirroring the ubiquity of the personal computer or the petroleum-powered automobile: a globally corporatized transition. The race to monopolize computer operating systems by private interests in the 1980s – for instance, the competition between Unix and MS-DOS – laid the foundations of the licensed computer ecosystem prevalent today, which includes 3D printing, although hybrid alternatives continue to capture markets, notably Google's Linux-based 'Android'. A similar guide for this quadrant is available in the ubiquity of the petrol-powered automobile. The influence of a consortium of petroleum lobbyists on the decline of the electric battery-powered vehicle that continues to hold sway in people's choice of vehicle demonstrates the lingering influence of business interests on consumer demand. In this world the realization of mass customization unleashes cultures of consumption reifying individualism and personal reinvention, further exacerbating the emotional costs and exuberances of globalization.[57] A major aspect of this world is the seamless nature of the 3D printing systems. Materials, designs, infrastructures and technologies are taken for granted by consumers in the same fashion as integrated assembly line factory production, transoceanic containerization, and brand marketing systems are today.[58] MNCs continue to dominate the material world in a post-mobilities setting of localized manufacturing.

The third quadrant's title is 'Sharevana' – a 3D society that is a nirvana for sharing, as it is collective and open. In this future the 3D printing ecosystem turns out to be wholly unsuitable for individual engagement with a combination of technological, design, infrastructural and material setbacks. Instead, the burgeoning demands for communal spaces for manufacturing manifest in an egalitarian correction in the global production, distribution and consumption triad. A governmental 'top down' push for nations 'carried aloft by the march of the makers' transforms public spaces and facilities into production hubs.[59] A model here is the UK Public Libraries Act 1850, which ushered in the liberal access of knowledge to 'one and all'.[60] Here too the diffusion of fablabs (fabrication laboratories) offers a window into the future.[61] Such 'makerspaces' envisage a democratic self-organization around urban peer-production: 'In the years to come, as fluidity and adaptability become the norm, urban dwellers will turn to their neighbourhood to leverage local connections, physical assets and the strengths of their peers'.[62] Projections of the future of peer-production propose a 'commons-oriented distributed manufacturing' as a tenable alternative to the global market economy driven by corporate capitalist interests.[63] Sharevana will certainly rely on a global commons in order to suppress monopolies and corporate co-optation of public assets.

'Photoshop' is the title for the fourth quadrant; a future world where manufacturing is localized and consumers enlist the services of spatially close clusters of production technology and expertise: corporations are close-and-personal with consumers in the mediation of 3D printing. The 3D printing ecosystem is neither compatible with public facilities nor is it available to individuals in their homes. Instead, digital fabrication is spurring a renaissance for post-industrial and newly

industrialized regions alike as spaces for reshored manufacturing without the perceived social ills of the current triad. Manufacturing decouples from distribution
and instead becomes a part of the consumption 'services' system. There is a further
'servicization' of the global economy: 'local, bespoke manufacturing-on-demand
based on 3D printing and a move to product plus service commercial models'.[64] In
the 'Photoshop' scenario 3D printers run in the background – that is, behind the
scenes – similar to containerization in today's system.

A model for this world is the ubiquity of the digital camera. Photographic
printing shops and services continue to remain popular for producing physical
copies of digital images, even with competition from home paper printers and
digital displays. In this instance, not having access to a desktop paper printer does
not preclude the mass redundancy of analogue cameras and ubiquity of digital
photography. Rather, the digital photography ecosystem – involving personal
computers, cameras, storage media, batteries, desktop printers, printshops or
store-based automatic print machines, and so on – creates new opportunities for
existing systems and services such as the traditional high street print shop. One
aspect of this is the continuing demand for analogue photography as a luxury or
professional form of photography in comparison to digital photography, which is
about disposability and mobility rather than quality per se.[65]

## Conclusion

So in this chapter we have outlined a method of forecasting that gets away from the
idea that futures depend on or are determined by technologies, or that they are simply
derived from the ways in which the present is unfolding. In the next chapter we
extend the future focus of this book to different test cases of possible worlds where
3D printing is ubiquitous. By no means are the scenarios presented in this book pre-
dictive of the way the future is going. Instead, realistic, well-informed and imaginative
present day themes guide each one according to axes of uncertainty. More likely than
not, the futurity of today will be as outlandish and nuanced as the examples at the
beginning of this book from the 1920s and 1930s. Certainly surprises are in store. The
point of the scenarios in the next chapter is not to paint the most accurate pictures of
probable futures, but instead to guide thinkers, policymakers, citizens (and their
social practices), social institutions and social groups into informed choices and pathways in order to mitigate and manage uncertainties in society. In short, putting the
human into futures. More than this is beyond the scope of this book.

## Notes

1  M. Sheller and J. Urry, 'The New Mobilities Paradigm'. *Environment and Planning A* 38,
   no. 2 (2006): 207–26.
2  M. Sheller, 'Uneven Mobility Futures: A Foucauldian Approach'. *Mobilities* 11, no. 1
   (2016): 15–31, doi:10.1080/17450101.2015.1097038
3  P. Merriman, M. Sheller, K. Hannam, P. Adey and D. Bissell, *The Routledge Handbook of
   Mobilities*. London: Routledge, 2014. p. 183.

4  J. Urry, *Societies Beyond Oil: Oil Dregs and Social Futures*. London: Zed, 2013.

5  —. *Climate Change and Society*. Cambridge: Polity Press, 2011.

6  M. Linz, 'Scenarios for the Aviation Industry: A Delphi-Based Analysis for 2025'. *Journal of Air Transport Management* 22 (2012): 28–35, doi:10.1016/j.jairtraman.2012.01.006

7  K. Gada, 'Reducing Poverty through Personal Manufacturing', 2011 Stanford Social Innovation Review. Accessed 20 August, 2012. http://www.ssireview.org/blog/entry/reducing_poverty_through_personal_manufacturing

8  J. Vanian, 'Why Ford is Partnering with a Hot 3D Printing Startup', *Fortune*. New York: Time Inc., 2015. 23 June 2015. Accessed 24 November 2015. http://fortune.com/2015/06/23/ford-hot-startup-3d-printing/

9  S. Harris, 'New Technique Paves the Way for 3D-Printed Aircraft Wings', Centaur Communications Ltd, 2013. Accessed 24 November, 2015. http://www.theengineer.co.uk/aerospace/news/new-technique-paves-the-way-for-3d-printed-aircraft-wings/1016759.article#ixzz3sN5nY79v

10  M. Potstada, A. Parandian, D.K.R. Robinson and J. Zybura, 'An Alignment Approach for an Industry in the Making: Diginova and the Case of Digital Fabrication'. *Technological Forecasting and Social Change* 102 (2016): 182–92, doi:10.1016/j.techfore.2015.07.020. 191

11  T. Birtchnell, S. Savitzky and J. Urry, eds., *Cargomobilities: Moving Materials in a Global Age*. London: Routledge, 2015.

12  N. Castree, A. Rogers and R. Kitchin, *A Dictionary of Human Geography*. Oxford: OUP, 2013. p. 355.

13  Intellectual Property Office, *3D Printing: A Patent Overview*. Newport: UK Government, 2013.

14  J. Söderberg, 'Reproducing Wealth without Money, One 3D Printer at a Time: The Cunning of Instrumental Reason'. *Journal of Peer Production*, no. 4 (2014): 1–10.

15  J. West and G. Kuk, 'The Complementarity of Openness: How Makerbot Leveraged Thingiverse in 3D Printing'. *Technological Forecasting and Social Change* 102 (2016): 169–81, doi:10.1016/j.techfore.2015.07.025

16  J. Söderberg, 'Reproducing Wealth without Money, One 3D Printer at a Time: The Cunning of Instrumental Reason'. *Journal of Peer Production*, no. 4 (2014): 1-10.

17  G. Stemp-Morlock, 'Personal Fabrication'. *Technology*, no. 1 (2009): 2–3, doi:10.1145/1831407. 2

18  J. Moilanen, A. Daly, R. Lobato and D. Allen, 'Cultures of Sharing in 3D Printing: What Can We Learn from the Licence Choices of Thingiverse Users?'. *Journal of Peer Production*, no. 6 (2015): 1–9.

19  L.S. Osborn, 'Regulating Three-Dimensional Printing: The Converging Worlds of Bits and Atoms'. *San Diego Law Review* 51 (2014): 553–622.

20  R. Fordyce, 'Manufacturing Imaginaries: Neo-Nazis, Men's Rights Activists and 3D Printing'. *Journal of Peer Production*, no. 6 (2015).

21  B.T. Wittbrodt, A.G. Glover, J. Laureto, G.C. Anzalone, D. Oppliger, J.L. Irwin and J.M. Pearce, 'Life-Cycle Economic Analysis of Distributed Manufacturing with Open-Source 3-D Printers'. *Mechatronics* 23, no. 6 (2013): 713–26, doi:10.1016/j.mechatronics.2013.06.002

22  M. Mavri, 'Redesigning a Production Chain Based on 3D Printing Technology'. *Knowledge and Process Management* 22, no. 3 (2015): 141–47, doi:10.1002/kpm.1466

23  D.H. Brean, 'Asserting Patents to Combat Infringement Via 3D Printing: It's No "Use"'. 2012 Webb Law Firm. Accessed 23 August 2012. http://works.bepress.com/context/daniel_brean/article/1000/type/native/viewcontent

24  T. Rayna and L. Striukova, 'From Rapid Prototyping to Home Fabrication: How 3D Printing is Changing Business Model Innovation'. *Technological Forecasting and Social Change* 102 (2016): 214–24, doi:10.1016/j.techfore.2015.07.023

25  T. Rayna, L. Striukova and J. Darlington, 'Co-Creation and User Innovation: The Role of Online 3D Printing Platforms'. *Journal of Engineering and Technology Management* 37 (2015): 90–102, doi:10.1016/j.jengtecman.2015.07.002

26 C. Weller, R. Kleer and F.T. Piller, 'Economic Implications of 3D Printing: Market Structure Models in Light of Additive Manufacturing Revisited'. *International Journal of Production Economics* 164 (2015): 43–56, doi:10.1016/j.ijpe.2015.02.020

27 M. Appleyard, 'Corporate Responses to Online Music Piracy: Strategic Lessons for the Challenge of Additive Manufacturing'. *Business Horizons* 58, no. 1 (2015): 69–76, doi:10.1016/j.bushor.2014.09.007

28 J. Moilanen, A. Daly, R. Lobato and D. Allen, 'Cultures of Sharing in 3D Printing: What Can We Learn from the Licence Choices of Thingiverse Users?'. *Journal of Peer Production*, no. 6 (2015): 1–9.

29 L. Xin and Y.U. Xiang, 'Potential Challenges of 3D Printing Technology on Patent Enforcement and Considerations for Countermeasures in China'. *Journal of Intellectual Property Rights* 20, no. 3 (2015): 155–63.

30 C. Bosqué, 'What Are You Printing? Ambivalent Emancipation by 3D Printing'. *Rapid Prototyping Journal* 21, no. 5 (2015): null, doi:10.1108/RPJ-09-2014-0128

31 D. Mercer, 'Scenarios Made Easy'. *Long Range Planning* 28, no. 4 (1995): 81–6, doi:10.1016/0024-6301(95)00015-B

32 H. Son, 'The History of Western Futures Studies: An Exploration of the Intellectual Traditions and Three-Phase Periodization'. *Futures* 66 (2015): 120–37, doi:10.1016/j.futures.2014.12.013

33 Ibid.

34 C. Bezold, 'Alternative Futures for Communities'. *Futures* 31, no. 5 (1999): 465–73, doi:10.1016/S0016-3287(99)00006-3

35 M. Godet and F. Roubelat, 'Creating the Future: The Use and Misuse of Scenarios'. *Long Range Planning* 29, no. 2 (1996): 164–71, doi:10.1016/0024-6301(96)00004-0

36 D.R. Morgan, 'The Dialectic of Utopian Images of the Future within the Idea of Progress'. *Futures* 66 (2015): 106–19, doi:10.1016/j.futures.2015.01.004

37 F.W. Geels and W.A. Smit, 'Failed Technology Futures: Pitfalls and Lessons from a Historical Survey'. *Futures* 32, no. 9–10 (2000): 867–85, doi:10.1016/S0016-3287(00)00036-7

38 S. Schnaars and P. Ziamou, 'The Essentials of Scenario Writing'. *Business Horizons* 44, no. 4 (2001): 25–31, doi:10.1016/S0007-6813(01)80044-6

39 D. Banister and R. Hickman, 'Transport Futures: Thinking the Unthinkable'. *Transport Policy* 29 (2013): 283–93, doi:10.1016/j.tranpol.2012.07.005

40 A. Curry, T. Hodgson, R. Kelnar and A. Wilson, *Intelligent Infrastructure Futures: The Scenarios – Towards 2055*. London: Government Office of Science and Technology, 2006. p. 25.

41 J.F. Coates, 'Scenario Planning'. *Technological Forecasting and Social Change* 65 (2000): 115–23. p. 121.

42 M. Godet, 'The Art of Scenarios and Strategic Planning: Tools and Pitfalls'. *Technological Forecasting and Social Change* 65, no. 1 (2000): 3–22, doi:10.1016/S0040-1625(99)00120-1

43 P.W.F. van Notten, J. Rotmans, M.B.A. van Asselt and D.S. Rothman, 'An Updated Scenario Typology'. *Futures* 35, no. 5 (2003): 423–43, doi:10.1016/S0016-3287(02)00090-3

44 W. Schultz, 'Models and Methods in Motion: Declining the Dogma Dance'. *Futures* 42, no. 2 (2010): 174–6, doi:10.1016/j.futures.2009.09.011

45 G.S. Kass, R.F. Shaw, T. Tew and D.W. Macdonald, 'Securing the Future of the Natural Environment: Using Scenarios to Anticipate Challenges to Biodiversity, Landscapes and Public Engagement with Nature'. *Journal of Applied Ecology* 48, no. 6 (2011): 1518–26, doi:10.1111/j.1365-2664.2011.02055.x. 1519

46 D. Bell, *The Coming of Post-Industrial Society: A Venture in Social Forecasting*. London: Heinemann, 1974.

47 Ibid., 9.

48 P.N. Stearns and D. Bell, 'Is There a Post-Industrial Society?'. *Society* 11, no. 4 (1974): 10–22, doi:10.1007/BF02701813

49 T. Birtchnell, G. Viry and J. Urry, 'Elite Formation in the Third Industrial Revolution'. In *Elite Mobilities*, edited by Birtchnell and Caletrío, 62–77. Abingdon: Routledge, 2013.

50  D. Guttridge, '3D Printing Mergers: 3D Systems Vs. Stratasys', Kapitall Inc., 2012. Accessed 11 August 2015. http://wire.kapitall.com/investment-idea/3d-printing-mergers-3d-systems-vs-stratasys/

51  V. Kostakis, M. Fountouklis and W. Drechsler, 'Peer Production and Desktop Manufacturing: The Case of the Helix_T Wind Turbine Project'. *Science, Technology & Human Values* 38, no. 6 (2013): 773–800, doi:10.1177/0162243913493676

52  K. Clay, '3D Printing Company Makerbot Acquired in $604 Million Deal', *Forbes*, 2013. Accessed 11 August 2015. http://www.forbes.com/sites/kellyclay/2013/06/19/3d-printing-company-makerbot-acquired-in-604-million-deal/

53  M. Holweg, 'The Limits of 3D Printing', *Harvard Business Review*. Boston, MA: Harvard Business Publishing, 2015. Accessed 4 August 2015, https://hbr.org/2015/06/the-limits-of-3d-printing

54  R. Hague, 'Richard Hague Questions Whether Everyone Will Have a 3D Printer', Science Museum, 2013. Accessed 11 August 2015, https://youtu.be/uSgpJesE-X0

55  L. Harouni, 'Lisa Harouni Thinks 3D Printing Can Make Design Accessible to You', Science Museum, 2014. Accessed 8 October 2013. http://youtu.be/0p0TmFlpc0E

56  C. Bosqué, 'What Are You Printing? Ambivalent Emancipation by 3D Printing'. *Rapid Prototyping Journal* 21, no. 5 (2015): null, doi:10.1108/RPJ-09-2014-0128

57  A. Elliott and C.C. Lemert, *The New Individualism*. Abingdon: Routledge, 2005.

58  T. Birchnell and J. Urry, 'The Mobilities and Post-Mobilities of Cargo'. *Consumption Markets & Culture* 18, no. 1 (2015): 25–38, doi:10.1080/10253866.2014.899214

59  G. Osborne, '2011 Budget: Britain Open for Business', Foreign and Commonwealth Office, 2011. Accessed 12 August 2015. https://www.gov.uk/government/news/2011-budget-britain-open-for-business

60  M. Featherstone, 'Archiving Cultures'. *The British Journal of Sociology* 51, no. 1 (2000): 161–84, doi:10.1111/j.1468-4446.2000.00161.x. 168

61  N. Gershenfeld, *Fab: The Coming Revolution on Your Desktop – from Personal Computers to Personal Fabrication*. New York: Basic Books, 2005.

62  L. Gansky, 'Interdependence: A Manifesto for Our Urban Future, Together'. *Architectural Design* 85, no. 4 (2015): 80–3, doi:10.1002/ad.1930

63  V. Kostakis and M. Bauwens, *Network Society and Future Scenarios for a Collaborative Economy*. New York: Palgrave Macmillan, 2014.

64  Foresight, 'Technology and Innovation Futures: UK Growth Opportunities for the 2020s'. London: The Goverment Office for Science, 2010. p. 6.

65  E. Shove and G. Walker, 'Caution! Transitions Ahead: Politics, Practice, and Sustainable Transition Management'. *Environment and Planning A* 39, no. 4 (2007): 763–70, doi:10.1068/a39310

# 7

# SCENARIOS

## Print-it-yourself

The scenario 'Print-It-Yourself' manifests from attention in the media to polymer extrusion (FDM) technologies and the potential of industrial additive manufacturing for 'a programmable personal fabricator ... able to make anything'.[1] Individuals engage with their own home or portable printers in their everyday routines, largely replacing the need for retail shopping for objects. The scenario is generally 'open' and corporate influence in personal 3D printing is minimal. The peer-to-peer sharing of 3D printer designs allows the technology to feasibly spread without any engagement in consumer markets.[2]

The sophistication of mass adopted 3D printer technologies is not as advanced as in other scenarios, principally because innovation occurs through grassroots collaboration rather than industry investment. In this scenario a range of 3D technologies in the digitally networked environment – scanners, printers, laptops, smart phones, modems – are ubiquitously synchronized with peers in the commons, enabling a socio-economic system of 'peer-production' engaged with across many homes and by individuals.[3] The do-it-yourself logic of this scenario extends to individuals' engagement with a range of informal suppliers of bulk printer parts and resources in order to facilitate for the popular participation in competitive open cultures of making.

This scenario involves ubiquitous household desktop printers assembled as a kit or bought fully assembled, or even printed pre-assembled or in parts in a self-replicating process from open source designs. Individuals use these to produce many or all of the objects they require or desire from freely shared peer-to-peer services and possibly even pirated online collections of designs, alongside formal online design retailers. Piracy is not a primary feature of this world due to the difficulty owners of intellectual property and authorities have with reining in widespread openness:

creators and inventors simply elect to make their innovations open source due to this being inevitable anyway.

The limitations of this scenario are in foreseeing the ubiquity of personal or home laser sintering or electron-beam melting. These printers are challenging today and innovations in supply and safety will necessitate upheaval in the same fashion as gun control in the US. If designs are freely available to individuals there will need to be systems of surveillance and control in place once the material and technical limitations are overcome.

In this scenario users engage with designs collaboratively, both reverse engineering existing ones and sharing their creativity with others. The development of intuitive software culminates in users tailoring stock designs to their own ends from templates. Design skills are a prerogative of education systems in this future and 3D printing impacts upon childhood learning and family life.[4] In this scenario there is a widespread familiarity with 3D CAD modelling skills from exposure to 3D printing in early education. The future is by all accounts a 'networked lifeworld' with a 3D ecosystem 'encircling the person'.[5] We forecast that there are competing demands for the benefits of openness in the home, following the trajectory of the Internet's current (although by some accounts threatened) availability and accessibility.[6] Children lead the adoption of 3D printing through curiosity as well as through competition with peers and adults through school assignments, activities and technical classes.[7]

A persistent problem in this world is design piracy following the proliferation of peer-to-peer sharing online. For ubiquitous 3D printing in such unregulated settings as the home there are many opportunities for patent violation. Being a world where open access prevails, alternatives tested by users themselves (and readily available) limit any systemic impacts of piracy in favour of design democratization.

Indeed it was a similar home revolution in peer-to-peer (P2P) websites and transferrable digital music files which had major consequences for the global music industry. These challenges emerged not from the corporate sector, but from small start-ups founded by young users, sometimes at school: Shawn Fanning, co-founder of Napster, who initially built the site to share his music with close friends around the country, is a case in point. For 3D printing the unregulated sharing of digital files printable as physical objects is a strong possibility.

Demand for travel by consumers for shopping has been curtailed by desktop printing or augmented to luxury 'unprintable' complex and electronic objects. Mass manufacturers of cheap and disposable products have been forced to move up the value chain. Infrastructure for the movement of finished products is being replaced by a competing market of feedstock suppliers, all trying to compel consumers to purchase their products. This change has led to greater standardization and automation of freight so that consumer feedstock supplies never run low.

The initial reaction to this scenario in the workshop was: how could global companies allow this to happen, given the potential disruptive threat it poses for current production and distribution companies? This scenario is predicated on low levels of 3D corporatization and participants felt that businesses would seek to re-establish control over the consumption and repair of their products in the home, with the result that corporatization would reassert itself. 3D printing will result in a shift from digital rights management embedded within designs, analogous to the shifts following the digitization of music and subsequent attempts by music companies to control its distribution.

A recent example of a sustainable community system is 'Willow Pond', a decentralized, low-carbon future residential community.[8] Drawing on trends for a typical suburban 'sprawl' residential development, a key trend in this exercise is the convergence of advanced technologies supporting decentralized production, in particular 3D printing. In 'Willow Pond' recycling becomes a major factor in the sustainability of this decentralized model, with technologies (assemblers/disassemblers) making materials reusable, depending upon the rate of decay.

The prototypes also played the role of introducing possible ideas about alternative business visions in this scenario. One idea raised in interviews with experts prior to the workshop was that of manufacturers of complex products, such as white-goods, making online repositories of parts available to their customers. These databases would carry downloadable designs, which product-owners could browse via identification numbers, and then replace modular non-critical parts themselves after printing these out at home. There are many issues with this however, such as the relative part strength of 3D printed objects, de-lamination due to the relatively weak layering process, and the downsides of 3D printing versus injection moulding.[9] Repositories would allow the diagnosis and addressing of simple issues most likely under warranty; more complex materials and electronics would still require repair by specialists.

While in-home 3D printing has disrupted global production, distribution and consumption, supply chains and distribution networks have remained intact due to the rapid growth in demand for powders and other printer feedstock. Print-It-Yourself describes a society of unlimited products available at the push of a button. The tactile physical world of consumer objects is developing the same way as digital audio and visual media. In this world, the standardized materials going into 3D printers are taken for granted – they are a 'god given right' – and the constant printing of objects is devaluing products, causing waste and raising thorny legal issues around insurance, safety and liability.

With printing on demand becoming a social norm, new practices around hygiene and convenience emerge. Increasing expectations for immediate gratification mean people are printing more objects and treating these as disposable. The unintended consequence here is the burgeoning demand for storage space and the side effect of exponential waste. However, there is a greater reuse and repair ethic than there used to be as more people repair broken objects by printing newer versions of the part. Recycling facilities are also being developed to help reduce the amount of waste and clutter in homes.

In addition, participants felt that consumers themselves might in this scenario start becoming more specialized in their use of 3D printers, seeking to capitalize on their production power – which again could shift towards greater corporatization and at the least marketization. Participants thought that beyond 2050 this scenario could result in a renaissance of (digital) cottage industries in the UK.

Emboldened by the lure of economic growth from a new innovation, planners shifted resources away from freight infrastructures, causing bottlenecks and choke-points to proliferate. Faced with disinvestment, many retail high streets lose customers through poor services. Domestic freight increases due to the success of small to medium-size online retailers. In addition, safety issues emerge in transport contexts as 3D printed parts do not match expectations and this causes disruption. Niche users are also engaging with 3D printing in unforeseen and unintended ways by illegally printing weapons, customized vehicle parts, drug factories, and black market goods. These users cause more unpredictable criminal and terrorist activities, thereby disrupting travel and transport planning further.

Participants identified a tension in the scenario between the greater reuse and repair ethic and the tendency for consumers to generate waste through unnecessary printing. Furthermore, the participants wanted to know how to counter the current culture of non-repair that currently existed. They suggested that an understanding of the future of planned obsolescence in the lifespans of products would give this scenario greater depth. In addition participants suggested that recycling and reuse needed to be separated to a greater extent in the scenario.

A potential consequence of this scenario for participants were issues of authenticity and value in 3D printing in the home: what value does an object have if it can be produced in the home and is consequently no longer scarce? Participants felt that the antiques market, for example, would become more valuable given the aura of antique furniture. Printed objects would then become less valuable than those that were difficult or illegal to print, that were authentic.

Finally, participants were keen to discuss issues of consumer convenience, asking whether 3D printing would develop along the lines of the division between home cooking and shop-bought ready meals. For example, they imagined that people in this scenario would design and print something if they could be bothered to or had the time and skills – and if they did not, then would they buy the product from a retailer with long supply chains?

## I print therefore I am

Even though many domestic, consumer-level 3D printers were developed, the printed products were judged to be of too low quality and the printers too technical to use and run, rather like the problems with videoconferencing around 2000. In the second scenario, 'I Print Therefore I Am', individuals engage with the 3D ecosystem – that is, scanning, designing and printing prototypes of objects – through mediators. Third party companies manage the actual printing process and mitigate the uncertainties and pitfalls of technology implementation and adoption. Consumers do not get their hands dirty, so to speak. Corporations in this scenario become truly 'lean' through decoupling their activities entirely from the physical world, becoming virtual entities. Consumers are empowered with distributed 3D printing technologies, yet there is much social inequality in access to technology based on income, status and power.

The surge of market economies around the 3D ecosystem is a result of significant upheavals in the interlocking global triad of production, distribution, and consumption. The corporatized digitization of these systems is in a sense playing catch-up with trends in global stock markets and finance in the late twentieth century. The rapid ubiquity and consequent sophistication of paper desktop printers illustrates the role of competitive businesses in technological advancement for 3D printing.

Consumers continue to harbour confusion about product life cycles and waste due to the complexity of product design and the incipiency of the market for secondhand items.[10] The disposal of apparently 'out of date' 3D printers is routine under the influence of brand marketing and innovations in planned obsolescence.[11]

A major hurdle in a personal 3D printer able to manufacture a true range of objects is the printing of metals in room, rather than laboratory, conditions. Notwithstanding on-going research and development with lasers, which is making headway in this future, this kind of printing is currently limited to low-melting point metals or metal-containing inks, both of which are problematic in terms of strength, conductivity and corrosion.[12] A viable option for a low-cost open-source metal 3D printer is a gas metal arc welder. A printer of this kind is capable of printing both aluminium and steel alloys and is inexpensive and safe for the home, involving a stationary weld gun as a print head.[13] The popular ownership of home welding and soldering kits is an intimation of this future, with units easy to procure, but requiring expertise to handle.

In this scenario individual engagement with design heuristics takes place in a global consumer culture privileging access to the most cutting-edge and novel 3D technologies and materials, which are owned and managed by corporations and other for-profit interests. The print on-demand business model means designs cater for individual consumers who are willing to pay a bit more for a unique product. The service costs are offset by individuals managing their own design choices and processes.

The integration of 3D printed body parts, functions and activities moves out of the medical sphere and into consumer markets.[14] The tailoring of objects to individuals' desires and wants results in a further marketization of body modification, in a similar manner to the cosmetic surgery industry.[15] Celebrities promote the personalization of 3D printed objects and their conflation with the mainstream fashion industry.[16] Design houses catering for luxury consumers affiliate customization with individual access to the most expensive 3D printer technologies. The latest smart materials allow designs to factor in change or growth over time, so-called four-dimensional printing: 'advances in additive manufacturing allow specific materials that possess controlled functional performance to be precisely deposited within a 3D space'.[17] Luxury individual engagement through the services of elite designers will allow previously unforeseen geometries and new horizons in design duplication to match the trademarked aesthetic details already present in elite luxury consumer markets for unique watches, handbags and haute couture.[18]

There are efforts to limit piracy by controlling supplies of materials. According to a corporate supplier we spoke with, one of the largest 3D printer companies for office and industrial models runs a recycling service from a single corporate site in the US where it restocks all of its customers' cartridges via its own facilities and supply chains. Many users have sought to transgress such supply chain monopolization. Another 3D printing community user we interviewed noted how they had managed to successfully hack their obsolete industrial printer, which was a model no longer supported by the manufacturer, in order to use their own refilled cartridges and continue to use the device without upgrading. Illicit suppliers of cartridges could proliferate in future.

The 'I Print Therefore I Am' scenario presents a vision of a ubiquitous 3D ecosystem emerging from a corporate infrastructure that would be much the same as conventional markets today which are increasingly online and digital. There is high individual engagement from consumer citizens who invest their identities in consumer objects of unparalleled uniqueness and differentiation. Brands and logos reign supreme within strict regimes that manage knowledge and product personalization, foretold by international treaties – a precursor being the Agreement on Trade Related Aspects of Intellectual Property Rights (TRIPs).[19]

Manufacturing infrastructures in this world are no longer centralized around bulk production. Corporations have now achieved their aspirations for digital 'leanness' through disengagement with the physical world of production. Companies are now 'design intensive' in order to adapt to ubiquitous distributed 3D printing.[20] MNCs are now providers of virtual objects in online marketplaces regulated by strict intellectual property guidelines and protected by globally recognized patents.

Consumer, retail infrastructures are more than ever investing in online products and services and are decoupled from global production networks and associated friction.[21] Digital brand outlets offer all manner of products and services driven by many start-up companies in competition for their own market niche. Consumption is easeful and streamlined, its leanness matching production and distribution. Consumers are moving just as much locally, or more, as they continue to enjoy investing their identities in consumer products that are now affordably customizable and bespoke.

Distribution infrastructure, chiefly containerization, is also leaner due to the shift to bulk handling of resources within 'cartridge economies'. Transoceanic freight systems have been retrofitted to manage 3D printing filaments, powders, liquids and so on. The geospatial technological revolution in logistics of the early twenty-first century now applies to the containerization of 3D printer feedstocks, which are regulated through 'security cultures' of precision and predictability.[22] Efficiency gains in product packaging and handling are combined with the exaggerated costs of 'official' printer cartridges, with importation and exportation strictly enforced by border forces and trade agreements.[23]

Rising incomes and standards of living in what are today manufacturing regions diminish 'comparative advantages' as new cohorts of consumers demanding their own markets level out 'imperfections' (some would say inequalities) in the infrastructure of global trade. 3D printing is a key facet in the levelling of the world economy. Decentralized, distributed production technologies destabilize global production networks and geographies of production.

Social movements targeting mass produced objects will provoke 'nostalgic' social practices and the opting out of customization cultures and 3D printing. So-called 'normcore' is an example – albeit a fringe one – of consumers rejecting boutique, limited edition, craft and unique objects in favour of mainstream, disposable and mass manufactured ones.[24]

People do not so much buy objects as pay for accessing or gaining a license to produce or download the design of one or more objects. This would be part of a growing 'access', rather than 'ownership', economy that has developed with the growth of the digital 'Internet of things' whereby objects are able to download designs based on their own artificial intelligence.[25] There is already a vast online open source network of designs and blueprints available for download and this could be matched to consumers via 'smart' 3D printers.

'Reverse Ludditism' could disrupt this world where a backlash against personal production, craft and individual making could increase the demand for mass manufactured objects. Also alternative social practices might take hold where virtual identities supersede the need for physical expressions of fashion, choice and diversity of products.

If people can print their own objects and clothes then they might choose simplicity and develop their identities offline rather than online: this backlash would be catastrophic for business interests in this world. Cyclic economies might be important for communities, especially if materials can be recycled and reused by consumers themselves on atomizers in localized or shared facilities.

Additive manufacturing remains a high-end, technically challenging process for producing objects that is only undertaken by small numbers of experts. Consumers interact with experts through their own advanced home rapid prototyping printers assisted by intuitive interfaces.

The neoliberal free 'capitalist' market is less organized than ever due to three causes that were predicted in the 1980s by one of the authors.[26] The scenario here is driven by changes in technology, for example in the 'microelectronics machine-tool industries': 3D printing is one in a suite of personal, distributed production tools available to consumers within a more disorganized system than today. Changes in taste also drive disruption in this world, particularly in the 'widespread rejection of mass consumption patterns' and heightened demand for individually distinct products. In place of bulk production individual (or bespoke) production takes precedence, facilitated by both digitally connected craft experts in niche industries and more mainstream 'brand' suppliers utilizing ubiquitous distributed production tools and online marketplaces.[27] Finally, there is competition on grounds of product quality.

3D printers are themselves consumer objects in this scenario and access across social classes is a source of dispute, protest and lobbying. Would 3D printing emerge as an urban or rural phenomenon? It could be that rural communities pioneer innovation as they are already experimenting with localism and close community interaction. Network technologies would allow disparate communities of users to troubleshoot and liaise virtually without costly expenditures on travel. Face-to-face connections are likely to occur on a local level, perhaps bringing together people from different backgrounds.

## Sharevana

In 2014 a bevy of laser sintering patents expired, opening the way for open source industrial 3D printers for community use. In the scenario 'Sharevana' people tend not to have 3D printers in their own homes due to challenges in skills, safety, cost, or simply time. Instead, informal, peer-support community networks of people participate in designing and crafting 3D products themselves in systems of 'peer-production'. What motivates these communities is the desire to collaborate on products and parts, partly in order to opt out of the global mass consumer market.

Sharevana describes a world where peer production and maker movements find favour over for-profit manufacturing. 3D printing continues to be too technical for most individuals to engage with on their own, instead community hubs are facilitating technology and personnel for all sorts of small-scale projects.

In this world the technology is not distributed in people's homes but rather appears from centralized service providers operating according to open access principles. It could be expected that public transportation will play an important role in connecting users with community resources. As well, facilities for the distribution of feedstocks will be managed by central authorities, in the same fashion as state-run utilities, either by state employees or sub-contractors.

In this scenario consumers become producers through common 'open' access to digital product designs. By sharing these via local 3D printing technologies with industrial capabilities – that is, laser sintering or electron beam melting – consumers-cum-producers are in a position to compete with established production firms, in short 'democratizing manufacturing'.[28] The widespread use of open-source and self-replicating machines, as well as easy access to the community craft centres, mean that business print shops and suppliers have little market success.

Many disposable and low quality items are 3D printed for minimal cost locally – beyond the raw materials and energy, subsidized by the state – from Massive Online Open Repositories (MOORs) of 3D design files. Sharevana is defined by much freedom but at the cost of a lack of standards and quality control. Services are unrefined and managed by volunteers, peers and generalists, with limited resources for controls and standards. The democratization of 3D printing in this scenario depends on how accessible and intuitive the interfaces are, and the expertise of the mediators – who will most likely be low paid or volunteers – for all society's citizens.

In this scenario standards are an important feature of how people make things: users rather than corporations bear the risk for faulty or otherwise unsafe object designs. If everyone produces the things they use according to their own judgements rather than mediators, it is unclear where the regulation will come from. Leadership and govern-ance are also issues in this scenario. In addition, different trends in consumer fashion could converge. A desire for 1970s 'vintage' designs could be met by printing out scanned or reverse engineered copies. Also, could too much choice be an issue? Instant gratification and print-to-demand might be the new way of 'doing' retail.

In Sharevana it is libraries, community centres and government initiatives which substitute for corporatized bureaus. Public facilities facilitate services and support for collective engagement with the 3D printing ecosystem. Instead of engaging directly with 3D printers, users interact with institutions to procure objects. These are designed and printed under supervision and with formal and informal technical assistance.

An important aspect of Sharevana's infrastructures is the co-creation and collabo-ration that occurs between people forming collectives and groups not only to man-ufacture objects but also to distribute and sell designs to wider groups of people. In the same manner as organic food and craft movements, the distribution of products occurs locally and items are traded or swapped communally, without the involve-ment of large corporations and perhaps without currencies (or with digital curren-cies). The costs of products will be based upon the materials used and the skill invested in design, rather than brand or rarity. Informal and tacit networks of users and small-scale suppliers challenge professionally run supply chains and logistics.

In this scenario locally based 'maker fairs' are popular locations for 3D printed products, where people go to exchange and barter their unique things. Indeed informal markets combining new technology with traditional practices are a marked response to dehumanized, mechanized, digitalized and formalized mass manufac-turing systems.[29]

The State intervenes in the materials economy means that fluctuations in resources and markets will not negatively impact government investments in the community-centred and user-led manufacturing model. Waste is not an option in this future where a shared responsibility for resources prevails and sophisticated systems manage supplies and match them to recycling and disposal.

One possible direction for feedstocks is closed-loop processes or 'circular econo-mies'. Currently, powders and other composite feedstocks have been derived from, for instance, recycled glass powders or other patent-protected resins: companies such as Z Corporation use their own 'non-hazardous, eco-friendly' powders. One solution that raises the possibility of a circular economy is a 'recyclebot' where plastic wastes, including old 3D 'prints', produce more feedstocks for new products in the home from a machine that accompanies or is built into the Makerbot 3D printer. Currently there are examples of circular economies utilizing waste-pickers in Pune, India, and 3D printers.[30]

Localism and sustainability agendas are cutting into corporate profits as self-organizing community movements are banding together against exploitative labour conditions in China and other manufacturing countries. Recent community energy cooperatives in the Netherlands are a step in this direction.[31] A new freight infra-structure, much like the petroleum infrastructure, has arisen for refined feedstock materials for printers. Governments sympathetic to equal access to 3D printing heavily subsidize the system and encourage ideologies of 'objects for all'.

Disposable packaging could become a resource in itself. Instead of being thrown away, litter could be added to a converter that would process this and transform it into the feedstock of a domestic 3D printer. Even more dramatically new social practices might render continuous supplies of raw materials redundant. Instead of having a surplus cache of consumer items 'to hand' for specific purposes, items could be printed when they are needed. Rather than using a plastic fork and then washing it up, the item could be 'recycled' as part of a hygienic, but possibly energy intensive, process into a spoon for dessert and so on. Clothes could be printed as needed and then recycled instead of cleaned as a new 'normal' way of laundering similar to how line drying was replaced by machine dryers.[32] Design plays a key role in these circular transitions and the rapid manufacturing of clothing promises zero-waste design practices, as the materials that go into clothing or other objects could be made compatible with the feedstock of 3D printers, which would be sustainably recycled using the same materials with lifetimes as long as the raw materials themselves.[33]

State support for the 3D ecosystem fosters the emergence of various community craft centres and shared 3D printing stations. People who hope to spur interest in open-source information and co-production lead these. The experts in our workshop wanted to know who would pay for this future. If it is neither corporate nor individual, it is unclear where the support of innovation comes from and how entrepreneurship thrives. Examples of fablabs and co-production or peer production networks are useful in clarifying this, as well as crowd sourcing and pledge drives, subscription models and these sorts of economies. Angel investors could kick-start innovators whose inventions might have mass appeal and ultimately State sponsorship. The transport implications of this scenario lead to highly routinized and planned services in order to bring users from their homes to community spaces. These could even be micro-timetabled according to online making/printing routines using digital alerts and portable devices.

## Photoshop

In the final scenario, 'Photoshop', 3D printing has met expectations for a renaissance of manufacturing in the Global North. The 3D ecosystem is the ultimate 'lean' system. There are no inventories, no transoceanic freight of finished goods, and consequently no import/export inequalities. MNCs no longer face challenging business relationships across regions.

Individuals engage with 3D printing in the high street and through suppliers, not by themselves. Many do not realize 3D printers are working behind the scenes. In this sense it is business-as-usual for global capitalism, which has perfected the just-in-time market economy for made objects.

Reshoring happens in many post-industrial countries. New market opportunities satisfy the demand for freight-less products. Many large MNCs abandon their global production networks entirely to become free floating entities, with no obligations to local governments beyond the storefront or website. There is however much economic growth for the Global North, as MNCs once again invest in local high streets, bureau systems and feedstock refineries. There is indeed a 'new industrial revolution' in this world.

A comparative study of the specific costs of material deposited in both EBM and SLS suggests that the observed deposition rates are insufficient for the integration of these metal printing technologies in high volume manufacturing applications, however the study also shows economies of scale are achievable in an alternative sociotechnical system where consumers demand 3D printed objects.[34]

In this scenario consumers might very well not notice they are engaging with 3D printing designs instead making use of services providing 3D printing to them. Local manufacturers will enlist crowd-sourced data in the design process, involving information procured directly from individuals through scanning and haptic interfaces.

An example of a major product in this scenario is wearable 3D prints. Consumers buying clothing will send their measurements from a three-dimensional scanner directly to a store and have the garment delivered right to their home once it has been 3D printed elsewhere.[35] The complexity of additively manufacturing garments and textiles combined with the necessity for assembly – that is, stitching or binding – means this kind of product is unlikely to be made in the home itself with no aid from an external party such as a fashion expert or textiles manufacturer. A possibility in this scenario is remote 'home delivery' information retrieval services: digital scanning could take place via agents who engage with consumers in their homes and offices for 'next day' printing services.

Many current products are very complex, and even the modest 'water bottle' is made up of various components. There will be design innovation in order for objects to be manufactured on-site using local manufacturing systems. Some assembly is probable. Therefore, employment opportunities will become available in the Global North for assemblers and other niche sectors.

Designs in this scenario will be the intellectual capital of various institutions whose role is to provoke wealth creation through the materialization of their knowledge assets. Since this scenario involves retailers managing the lion's share of 3D printing, the designs will be an invisible aspect of the process for consumers who will instead engage with customizable templates – generic objects with a range of features with customizable potential.

Retail stores are innovating infrastructure by sharing facilities and training their staff in customizing a wide range of generic consumer products, as well as using 3D scanners for in-house object tailoring. Because market innovators print only what they need, usually on site, global production networks cannot compete through lean supply chains, and their inefficiencies are made worse by the appearance of local printshops and bureaus in China, India, and other manufacturing countries. These local competitors satisfy their citizens' demands for consumer goods and middle-class lifestyles, pushing up salaries as consumption increases, and flattening out international income inequalities.

Transport and energy use are affected by gains in efficiency at all stages of the supply chain led by new business models and corporate investments in distributed networks of high-end printers. Some businesses are now dealing directly with suppliers of resources who adapt their facilities to deliver materials in forms that are compatible with printers.

There would be a monopolization of feedstocks and even a new patent feedstock (the experts in the workshop suggested the name 'Printium' or alternatively 'Unobtanium'). Access to materials for printing will become crucial for consumer societies to function, much like oil for car cultures today wherein citizens cannot imagine doing without petroleum. The heightened mobility of the early twenty-first century will face constraints from competing demands for resources in transport. Consumers will question the 'burning' of gas and oil for transport and lobby instead for affordable access to materials for object production, in the same way as the cost of 2D printer inks inhibits some printing of paper documents.

One issue for this scenario, in regard to materials, is scale. Feedstocks of materials will be purchased in bulk and distribution networks will operate using retrofitted existing supply chains, heavily standardized for 3D printer cartridges. Inventories will be near to or in retail hubs as time is crucial for physical stores, though less so for virtual 'online' retailers.

It is no easy task to produce a simple product with a 3D printer even by today's standards. Many products in the contemporary world are multi-component and require assembly, mixed materials and electrical conductivity. Recycling was discussed and whether the recycling of printed products could turn those into an array of feedstocks. The conditions whereby the current global transportation and production system might be disrupted were also raised. The closure of the Suez Canal and massive increases in the costs of shipping were examples of possible significant disruptions to existing fossil fuel or delivery systems. Does Moore's Law apply for 3D printing and if so does this mean there will be a similar trajectory of exponential development and progress?

Experts in the workshop discussed how people, while not engaging first-hand with the design and printing process, will access local bureaus or 'print-shops' for printing objects: a booming new economy for post-industrial 'services'. These services represent a blossoming of regional and local manufacturing, which is corporatized and composed of markets for designs, material feedstocks and 3D printer technologies. Containers full of manufactured objects are a thing of the past, and much of the logistics of production and delivery have been retrofitted or replaced by flows of resources for local printing.

Due to the scope for efficiency and inventory gains from 3D printing, MNCs meet their emissions targets for GHGs by reducing the international freight of finished products. In place are 'smart' domestic freight systems, which move 3D printed objects through 'print-on-demand' business models. Much distribution is automated through innovations including land, sea and air drones (uncrewed 'automated' vehicles) of differing sizes. Key to this domestic freight are door-to-door delivery services and the mail system.

With in-home 3D printing failing to develop alongside the growing trend of people shopping online, the growth of advanced manufacturing technologies allows companies to invest in and promote complex, novel and technical objects mediated by their own employees. In doing so 3D printing is to a point monopolized and restrictions are probable over the 'bootlegging' of object production.

Additive manufacturing will blend with existing manufacturing techniques to a degree, and many companies will elect to undertake smaller runs of products and parts through as an alternative to inventories and long-distance freight. Companies will be the main beneficiaries of 3D printing and will integrate this into their business models and systems.

3D printing will become a core aspect of production as the container did in the shift to 'containerization'. The innovation will be managed within corporate structures and operated by trained experts with qualifications and skills. It will not develop as anything like a new system, but instead will add to the range of current manufacturing affordances within smaller-scale factories and workshops, different from but also similar to today.

## Futures beyond mass as we know it

In this book we have intended to show how the ubiquity of 3D printing will be more than anything else of profound social consequence, and we have also argued that there are four possible scenarios for 2050 as way stations for different transition pathways. A key notion in all of these is that people will integrate the processes involved rapidly and in innovative ways that are unimaginable today. In the first, Print-It-Yourself, the public commons expands to overtake neoliberalism as openness prevails and individuals engage singly with 3D printing in a decentralized manner. On the whole, the emergence of the Internet offers a clue to the flavour of this scenario as individuals compete to utilize freely available 3D printing designs and technologies. The economic consequences of mass individual engagement with open innovations are challenging to fathom, with market systems neither being profitable nor essential for object procurement. There is a pattern of *resource centrism* in this world, as expertise no longer adds value to objects that individuals are able to print themselves with little to no cost or mediation from others. While the means of production is freely available to individuals, resources are not. Access to, and ownership of, resources for 3D printing will usher in new dimensions to geopolitics for nation-states with mineral wealth, and prehistoric models of trade in resources (that is, the Neolithic movement of tin from the UK or amber from the Baltic) will offer insights for this scenario.

In the second 'I Print Therefore I Am' scenario, there is an overwhelming sense of neoliberalism as the free market system continues to dominate the world's trade in objects, as ideas of individual meritocracy enmesh even further with material wealth. The cultures of consumerism prevalent today in people's corporeal and intimate activities – that is, cosmetic surgery, lifestyle tourism, life 'coaching' and recreational shopping – blend seamlessly with 3D printing's affordances. A rejection of 'mass' made objects is not inconceivable as a driver for 3D printing's ubiquity in this scenario: people want objects that represent their personalities, egos and desires. People 'live to print' in this world, and corporate profit no longer derives from the manufacturing and haulage of objects, but rather the service and supply to individuals of resources and intellectual property. Corporations control the knowledge invested in objects. This is *post-production*, as the lion's share of objects (notwithstanding the 3D printers themselves) are 'made' by individuals and not corporations, who instead focus their efforts on inciting consumer trends for the cyclic renewal of designs and innovative, 'smart', materials. Planned obsolescence intensifies in this world, as do fashion cycles and investment in cultures of excess. A cultural celebration for bespoke objects, individuals' genius and expertise, 'local' products, and aesthetic spectacles and creativity in modern life is gaining momentum (with parallels in the current corporatized 'hipster' movement today). Far from a rebellion against homogeneity, a renaissance in heterogeneity defines this world.

In the third scenario of 'Sharevana', 3D printing emerges as a collectively embraced technology aligned with cultures of sharing and openness, yet divorced from an individual capacity for object production. In this world there is a need for 'gurus' who innovate designs for objects and stipulate what and how groups should 3D print. Such individuals could be motivated by philanthropic urges or from ideologies akin to those found in some spiritual organizations or non-profit social movements. Object production and social interaction are combined in a *post-privatization and profit* pattern of society. Democratic or centralized socialist societies are governance models for this world wherein profit is not kept private but located in the commons. The means of production turns out to be a vital feature in the political motivation for individual self-organization and solidarity against the privatization of space and wealth.

In the fourth scenario of 'Photoshop', 3D printing merges seamlessly into the 'back end' of the current production–distribution–consumption triad, with the majority of individuals not even noticing 3D printers have become ubiquitous. What individuals do notice is the greatly increased speed of delivery, the lower cost of products, and the greater choice and variety available in everyday objects. The push for leanness in manufacturing and supply chains has resulted in an incredible degree of standardization in the transoceanic freighting of raw resources, in some cases moving through pipelines in the same fashion as utilities of water and gas in homes today. 3D printers are a feature of the *social fabric* in this world, with individuals not needing to 'get their hands dirty' as corporations manage the processes of production and distribution of 3D prints. Business models face the most change in this world, with corporate collectives determining how 3D printing is profitable and balancing its ubiquity with other fabrication options.

Across all four scenarios climate change is responded to differently. Sustainability in terms of societies' reliance on fossil fuels, with consequent GHG emissions from power and materials for printing, will differ dramatically. In a world of individual engagement and openness, as in 'Print-It-Yourself', there will be less standardization in designs, more misprints, and the prevalence of convenience as a personal choice.

These issues will cause some waste that could be balanced by people's willingness to be frugal and prudent. Policymakers will need to lobby citizen-consumers to print less, conserve materials, print only replaceable parts, and opt for recycling processes alongside printers – for instance personal 'shredder bots'. Off-gridders would be able to audit their own energy use and use of resources.

With the individual engagement and corporatization of 'I Print Therefore I Am' there will be few incentives for restraint and business-as-usual is likely in this high-carbon future. Waste will be profuse and corporations will require incentives to 'go green' at the expense of profit from fossil fuels. With consumer-citizens being driven to 3D print by profiteering corporations there is not much scope for a low-carbon transition.

In 'Sharevana' a society-wide response to climate change and a low-carbon system is feasible, however the governance of what people print and why will be counter-balanced with concerns about totalitarianism, control and surveillance. Policymakers would hold the key in radically promoting the use of renewable energies and enforc-ing the recycling of materials and the curation of 3D printing.

Finally, in 'Photoshop' the efficiency gains made from decoupling fossil fuels from transportation and even production could translate into far fewer GHG emis-sions and the ultimate 'lean' system of a circular economy. The awareness of object lifespans from cradle-to-grave, and the stewardship of objects by the corporations who brand them, is linked to profit in this world, and the inefficiencies and envi-ronmental costs of fossil fuels could lead corporations to disinvest in these without jeopardizing their 3D printing operations.

All four futures broadly stem from a reconciliatory approach to thinking on systems and social practices at multiple levels. Neither are 3D printers a technology fix for the many drivers of change understood as prompting change today, and nor are they utterly disruptive to the current triad of interlocking systems. Instead they are somewhere in between. They suggest both alternative and complementary social practices and at the same time either compatible or unsympathetic systemic nuances. What is apparent from a critical appraisal is that the ubiquity of 3D print-ing would be on a par with similar innovations: the automobile, the mobile phone, and the Internet.

Societies before prior technological upheavals were fundamentally different from how they are now and the same is inevitable for 3D printing should it reach ubiq-uity. Yet if social responses to these other innovations are anything to go by, there will also be much inertia, resistance, and continuation of current standards, practices and traditions. Path dependencies are complex in social transitions. The computer revolution indeed represents a radical shift in how people live their lives, although in hindsight this is challenging to perceive.

Currently one of the authors is writing this paragraph on a lightweight portable laptop (a mobile office) in a gully of sub-tropical rainforest filled with birdsong and the rustling of palm leaves in the forest canopy above. There is no Internet connec-tion, power source, or any other office (in)conveniences. He has at his fingertips copies of nearly all of the sources in this book's bibliography, a full *Oxford English*

*Dictionary* and thesaurus, a music library for entertainment and inspiration, and countless other features that could only be found in workplace environments half a century ago. Yet many workers in the knowledge economy or elsewhere would not be permitted such freedom and trust, and are still bound to physical locations for work (despite long commutes by private or public transport) and the same nine-to-five routine as before portable computers became ubiquitous. Transition is incremental and prone to politics, policy and institutional short-sightedness, anachronism and parochialism. Surely too, 3D printing will not reach ubiquity without ramifications for society.

Finally we can inquire across all worlds whether there is an alternative to massification throughout. Critical insight from communications studies on the massification of media suggests demassification is not a necessarily antediluvian optic in the future, as it is afforded by the latest technological innovations at a grassroots level: digital shopfronts, online trade networks and portable financial tools, and ICTs. A key paradox is that in order for production, distribution and consumption to demassify, there will need to be stepping stone technologies that depend on this system to reach ubiquity: 3D printers are one such example of 'mass' manufacturing continuing into the future.

No doubt this book will sit on bookshelves alongside other more ardent or fetishizing ones containing momentous predictions determined by the technological affordances of 3D printing. Hopefully we have offered a more measured and prosaic account of this technological innovation by reviewing a corpus of peer-reviewed and scholarly accounts from the social sciences, engineering and computer science. We have been at pains to balance our inquiry between the technical aspects and the social ones. It is tempting in a book on a technological process to be deterministic, and throughout this one we have tried to look beyond the instrumental advantages and disadvantages of 3D printing to the social, environmental and economic aspects and consequences. Indeed our central argument has been that 3D printers are not very proficient or scalable production tools in the current system of bulk volume mass manufacturing: they only make sense once the gaze is widened to foresee how people will use and apply them to various aspects and applications of society.

We have seen that there is no single future, but many that are related to various social uncertainties. The possibility of decentralizing and distributing the means of production is by no means trivial, and social scientists are now only beginning to acknowledge the many cumulative drivers of change towards the democratization of manufacturing. Yet we have shown there are many directions a transition could take and some will not be a shift away from consumer capitalism and its global reach.

## Notes

1 N. Gershenfeld, *Fab: The Coming Revolution on Your Desktop – from Personal Computers to Personal Fabrication*. New York: Basic Books, 2005. p. 4.

2  J. Moilanen and T. Vadén, '3D Printing Community and Emerging Practices of Peer Production'. *First Monday* 18, no. 8-5 (2013), doi:10.5210/fm.v18i8.4271

3  Y. Benkler and H. Nissenbaum, 'Commons-Based Peer Production and Virtue'. *Journal of Political Philosophy* 14, no. 4 (2006): 394–419, doi:10.1111/j.1467-9760.2006.00235.x

4  D. Barlex and M. Stevens, 'Making by Printing – Disruption Inside and Outside School?', 2012 PATT 26 Conference, Technology Education in the 21st Century, Stockholm, Sweden, Linköping University Electronic Press, Linköpings universitet 26–30 June 2012.

5  T. Rasmussen, *Personal Media and Everyday Life*. Basingstoke: Palgrave Macmillan, 2014.

6  E. Maxwell, 'Open Standards, Open Source, and Open Innovation: Harnessing the Benefits of Openness'. *Innovations: Technology, Governance, Globalization* 1, no. 3 (2006): 119–76, doi:10.1162/itgg.2006.1.3.119

7  M. Eisenberg, '3D Printing for Children: What to Build Next?'. *International Journal of Child-Computer Interaction* 1 (2013): 7–13, doi:10.1016/j.ijcci.2012.08.004

8  B. Tonn and D. Stiefel, 'Willow Pond: A Decentralized Low-Carbon Future Scenario'. *Futures* 58 (2014): 91–102, doi:10.1016/j.futures.2013.10.001

9  A. Dean, '3D Printing in the Home: Reality Check', 2012. Accessed 19 August 2012. http://develop3d.com/features/3d-printing-in-the-home-reality-check

10  A. Dindarian, A.A.P. Gibson and J. Quariguasi-Frota-Neto, 'Electronic Product Returns and Potential Reuse Opportunities: A Microwave Case Study in the United Kingdom'. *Journal of Cleaner Production* 32 (2012): 22–31, doi:10.1016/j.jclepro.2012.03.015

11  G. Slade, *Made to Break: Technology and Obsolescence in America*. Harvard: Harvard University Press, 2006.

12  C.W. Visser, R. Pohl, C. Sun, G.-W. Römer, B. Huis in 't Veld and D. Lohse, 'Toward 3D Printing of Pure Metals by Laser-Induced Forward Transfer'. *Advanced Materials* 27, no. 27 (2015): 4087–92, doi:10.1002/adma.201501058

13  A.S. Haselhuhn, B. Wijnen, G.C. Anzalone, P.G. Sanders and J.M. Pearce, 'In Situ Formation of Substrate Release Mechanisms for Gas Metal Arc Weld Metal 3-D Printing'. *Journal of Materials Processing Technology* 226 (2015): 50–9, doi:10.1016/j.jmatprotec. 2015.06.038

14  D. Lupton, 'Fabricated Data Bodies: Reflections on 3D Printed Digital Body Objects in Medical and Health Domains'. *Social Theory & Health* 13, no. 2 (2015): 99–115, doi:10.1057/sth.2015.3

15  A. Elliott, 'Drastic Plastic and the Global Electronic Economy'. *Society* 46, no. 4 (2009): 357–62, doi:10.1007/s12115-009-9226-5

16  —, '"I Want to Look Like That!": Cosmetic Surgery and Celebrity Culture'. *Cultural Sociology* 5, no. 4 (2011): 463–77, doi:10.1177/1749975510391583

17  E. Pei, '4D Printing - Revolution or Fad?'. *Assembly Automation* 34, no. 2 (2014): 123–27, doi:10.1108/AA-02-2014-014

18  J. Urry, 'Consuming the Planet to Excess'. *Theory, Culture & Society* 27, no. 2–3 (2010): 191–212, doi:10.1177/0263276409355999

19  B. Christophers, 'The Territorial Fix. Price, Power and Profit in the Geographies of Markets'. *Progress in Human Geography* 38, no. 6 (2014): 754–70, doi:10.1177/ 0309132513516176

20  C.J. Hatch, 'Geographies of Production: The Institutional Foundations of a Design-Intensive Manufacturing Strategy'. *Geography Compass* 8, no. 9 (2014): 677–89, doi:10.1111/gec3.12158

21  N.M. Coe and M. Hess, 'Global Production Networks, Labour and Development'. *Geoforum* 44 (2013): 4–9, doi:10.1016/j.geoforum.2012.08.003

22  A. Kanngieser, 'Tracking and Tracing: Geographies of Logistical Governance and Labouring Bodies'. *Environment and Planning D: Society and Space* 31, no. 4 (2013): 594–610, doi:10.1068/d24611

23  S.M. Tobias, 'No Refills: The Intellectual Property High Court Decision in Cannon v. Recycle Assist Will Negatively Impact the Printer Ink Cartridge Recycling Industry in Japan'. *Pacific Rim Law & Policy Journal* 16 (2007): 775–804.

24 D.R. Shah, 'Life after Normcore'. *Textile View*. Metropolitan Publishing, Autumn/Winter 2015. Accessed 31 August 2015. http://view-publications.com/textile-view-issue-109/

25 J. Rifkin, *The Third Industrial Revolution: How Lateral Power is Transforming Energy, the Economy, and the World*. Basingstoke: Palgrave Macmillan, 2011.

26 S. Lash and J. Urry, *The End of Organized Capitalism*. Cambridge: Polity Press, 1987. p. 199.

27 S. Luckman, 'Women's Micro-Entrepreneurial Homeworking'. *Australian Feminist Studies* 30, no. 84 (2015): 146–60, doi:10.1080/08164649.2015.1038117

28 C. Weller, R. Kleer and F.T. Piller, 'Economic Implications of 3D Printing: Market Structure Models in Light of Additive Manufacturing Revisited'. *International Journal of Production Economics* 164 (2015): 43 56, doi:10.1016/j.ijpe.2015.02.020. 55

29 N. Hansson and H. Brembeck, 'Market Hydraulics and Subjectivities in the "Wild": Circulations of the Flea Market'. *Culture Unbound: Journal of Current Cultural Research* 7 (2015): 91–121, doi:10.3384/cu.2000.1525.157191#sthash.AsPKmVH9.dpuf

30 T. Birtchnell and W. Hoyle, *3D Printing for Development in the Global South: The 3D4D Challenge*. Basingstoke: Palgrave Macmillan, 2014.

31 J.A.M. Hufen and J. Koppenjan, 'Local Renewable Energy Cooperatives: Revolution in Disguise?' *Energy, Sustainability and Society* 5, no. 1 (2015): 18.

32 E. Shove, 'Sustainability, System Innovation and the Laundry'. In *System Innovation and the Transition to Sustainability*, edited by Elzen, Geels and Green, 76–94. Cheltenham: Edward Elgar, 2004.

33 P. Delamore, 'Case Studies in Digital Textile Printing'. *Textile Forum*, no. 3 (2007): 30–1.

34 M. Baumers, P. Dickens, C. Tuck and R. Hague, 'The Cost of Additive Manufacturing: Machine Productivity, Economies of Scale and Technology-Push'. *Technological Forecasting and Social Change* 102 (2016): 193-201, doi:10.1016/j.techfore.2015.02.015

35 A. Beazley, 'Twenty-First Century "Made-to-Measure"'. *Costume* 38, no. 1 (2004): 116–25, doi:10.1179/cos.2004.38.1.116

# INDEX

For Product Safety Concerns and Information please contact our EU representative GPSR@taylorandfrancis.com Taylor & Francis Verlag GmbH, Kaufingerstraße 24, 80331 München, Germany

Printed and bound by CPI Group (UK) Ltd, Croydon, CR0 4YY
01/05/2025
01858552-0001